TYPOGRAPHY AND ELECTROTYPING BY
F. H. GILSON COMPANY
BOSTON, MASS.

PRESSWORK BY
BERWICK & SMITH CO.
NORWOOD, MASS.

PREFACE.

THE purpose of this volume is to present in compact form certain approximate methods of determining the true bearing of a line, together with the necessary rules and tables arranged in a simple manner so that they will be useful to the practical surveyor. It is a handbook rather than a text-book, hence many subjects have been wholly omitted which are ordinarily included in books on Practical Astronomy but which are not essential in learning to make the observations described in this book. In all of the methods here treated the object sought is to secure sufficient accuracy for the purpose of checking the measured angles of a survey with the least expenditure of time. For this reason many approximations have been made and many refinements omitted which simplify the calculations without introducing serious error into the results, and although such a treatment would scarcely be proper in a text-book the gain in simplicity and convenience would seem to justify its use in a book of this character.

The necessity for making astronomical observations for azimuth is confined chiefly to geodetic work, and arises so seldom in general engineering practice that many persons engaged in surveying are not familiar with astronomical methods and will not feel confident of obtaining reliable results, and therefore are likely to avoid making use of such observations even when they might be of great practical value. This consideration, together with the fact that the rigorous methods of calculating azimuth are rather long and complex, have tended to prevent astronomical observations from being more generally applied to surveying. The author has endeavored to so present the subject that a person who is unfamiliar with astronomy will be able to apply these methods and obtain satisfactory results without taking the time to completely master the theory underlying the method used. The rules and tables have been put in compact form so that the book may be carried in the field and the results of observations worked up at once if desired. The value of having, from time to time, an independent check on the angular measurements of an extensive survey will certainly warrant spending a few minutes' time in making observations and computing the results.

The methods here presented are not new but have all appeared in one form or another in works on Navigation, Astronomy, and Surveying.

iii

Much valuable matter written on this subject is so scattered, however, that it is difficult to find in one small book all that would be needed by the surveyor in making azimuth observations.

The author desires to acknowledge his indebtedness to Professor C. F. Allen for the use of the electrotype of Table XVIII taken from his "Field and Office Tables," and to Professor F. E. Turneaure for permission to reprint Table V from Johnson's Theory and Practice of Surveying. Thanks are due to Professor C. B. Breed for valuable suggestions and criticism of the manuscript.

G. L. H.

BOSTON, MASS., *January*, 1909.

CONTENTS.

Art.		Page
1.	Checking the Angles of a Survey.................	1
2.	Sun and Star Observations......................	1
3.	Apparent Motions of the Stars — Meridian.........	2
4.	Polar Distance — Declination...................	2
5.	Hour Angle...................................	3
6.	Latitude and Elevation of Pole................	3
7.	Correcting an Altitude — Refraction Correction....	3
8.	Index Correction..............................	3
9.	Making Solar Observations.....................	4
10.	Making Star Observations......................	4
11.	Azimuth Mark.................................	5
12.	Convergence of Meridians......................	5

Methods of Observing

13.	Azimuth by an Observed Altitude of the Sun........	6
14.	Computing the Azimuth.........................	8
15.	When to Observe..............................	10
16.	Azimuth by an Altitude of a Star................	13
17.	Observations for Latitude......................	13
18.	Latitude by the Sun at Noon...................	13
19.	Azimuth and Latitude by Observation on δ Cassiopeiæ and Polaris....................................	15
20.	Finding the Stars.............................	15
21.	Explanation of the Method.....................	15
22.	The Tables...................................	19
23.	Making the Observations.......................	20
24.	Observations on δ Draconis....................	23
25.	Meridian Line by Polaris at Culmination.........	25
26.	Azimuth by Polaris when the Time is Known........	26
27.	Accurate Determination of Azimuth by Polaris......	27
28.	Determining the Hour Angle....................	28
29.	Determining the Azimuth.......................	29
30.	Computing the Azimuth.........................	29
31.	Meridian by Polaris at Elongation...............	33
32.	Meridian by Equal Altitudes of a Star...........	35
33.	Meridian by Equal Altitudes of the Sun..........	37
	Tables.....................................	39–73

v

AZIMUTH.

1. Checking the Angles of a Survey by Astronomical Azimuths. — In the following pages are given several short and convenient methods of determining the azimuth of a line with an engineer's transit. While these methods may be used to determine an azimuth for any purpose which does not require great precision, the formulæ and the tables have been specially arranged so that it will be practicable to compute the azimuth in the field for the purpose of checking the angles of a survey. In a preliminary railroad survey, for instance, or in running long traverses by the stadia method, there will ordinarily be no reliable check on the measured angles such as that obtained by closing a circuit or by connecting the survey with some line of known azimuth. In such cases the angles can be checked, with all the accuracy required, by determining the azimuth of some line of the survey either by means of a sun observation or by an observation on the pole-star, and comparing the azimuth thus determined with the azimuth computed by means of the measured angles. With convenient tables the azimuth may be computed in the field in a few minutes' time, so that it will be known at once whether the preceding angles are correct.

The methods explained in Arts. 13 to 24 inclusive will give results sufficiently accurate for a check of the angles of an ordinary survey, but in these methods extreme accuracy is sacrificed to convenience and rapidity in the computations in order that results may be quickly obtained in the field. In Art. 27 is given a more accurate method which may be used when it is necessary to obtain an azimuth that is correct within a few seconds.

2. Sun and Star Observations. — With regard to the relative advantages of sun observations and observations on the pole-star it may be said that observations on the sun are the more convenient of the two, but will not give results of great accuracy; their particular advantage is that they can be made while the survey is in progress and with the loss of but a few minutes' time, and if the azimuth is desired only to about one minute of angle, sun observations will be sufficiently accurate. By observations on the pole-star the azimuth may be determined with great accuracy, and such observations are the ones most commonly employed when the best

results are sought. Star observations, however, have the disadvantage
that they must be made at night, in which case the surveyor has to make
a special trip to the point of observation, and must carry on his work
under various practical difficulties which are not encountered in making
sun observations. If a precise azimuth is desired it is necessary to observe
on the pole-star, or some other star close to the pole, and also to make
auxiliary observations for latitude and hour angle, on which the computed azimuth depends. The data for such an observation must be
obtained from the Nautical Almanac. If, however, only an approximate
result is desired, say within about 1 minute, the observations may be
made on the pole-star by the method given in Arts. 19 to 24, and the
azimuth found by Tables VII to XII. This method has the advantages
that no Nautical Almanac is required and that there is little calculation
except interpolation in the tables.

3. **Apparent Motions of the Stars — Meridian.** — If one watches the
stars for several hours he will observe that they all appear to move from
east to west in circular paths as though they were all attached to the
surface of a great sphere, and this sphere were turning about an axis,
the earth being at the centre of the sphere. Those stars which are near
the equator all move in large circles. As the observer looks farther north
he sees that these circles grow smaller, their common centre being an
imaginary point called the *pole*. A vertical plane through the pole cuts
out on the sphere a great circle called the *meridian* of the observer. The
line in which this meridian plane cuts a horizontal plane through the
observer is called the *meridian line*. At a distance of about 1° 10′ from
the north pole is a bright star which moves around the pole in a very small
circle and is known as the *pole-star*, or *Polaris*. Since its circle is small
its apparent motion is very slow, and consequently it is easy to determine
with accuracy its true bearing at any instant and to measure an angle
from the star to a reference mark on the ground.

The relative position of the fixed stars * is always practically the same,
and the pole stays nearly in the same position among the stars year after
year. We may think of all of the stars in the north as moving in circles
around the pole once each day, the size of the circle of any star and the
speed of the star's motion depending upon how far that star is from the
pole.

4. **Polar Distance — Declination.** — The angular distance to a star
from the pole is known as its *polar distance*, an angle which changes
slightly from year to year and may be obtained for any date from the

---

* The term *fixed star* is used to distinguish the very distant stars from the planets; the
latter are within the solar system and consequently appear to change their positions
rapidly. The fixed stars have but a slight motion, imperceptible to the naked eye.

American Ephemeris and Nautical Almanac.* In case of the sun or a star which is far from the pole it is more convenient to define its position by its angular distance from the equator, i.e., by the *declination*, which is the complement of the polar distance.

5. **Hour Angle.** — The *hour angle* of a body is the number of hours, minutes, and seconds that have elapsed since the body was on the observer's meridian. Hence it is simply the angle through which the body has appeared to move since it passed the meridian of the observer. It will be seen then that a star which has an hour angle between 0^h and 12^h is west of the meridian; if the hour angle is between 12^h and 24^h the star is east of the meridian. Hours or degrees are simply units of measurement of a circumference; in one case the circle is divided into 24 hours, and in the other case into 360 degrees. Hence an hour angle which is expressed in hours, minutes, and seconds of time may be converted into degrees, minutes, and seconds of arc. Since $24^h = 360°$ we have $1^h = 15°$, or, in a more convenient form, $1° = 4^m$ and $1' = 4^s$.

6. **Latitude and Elevation of Pole.** — A very important principle in astronomy is that the *angular altitude of the pole above an observer's horizon equals the latitude of the observer*. Hence the latitude may be determined by simply finding, by some means, the altitude of the pole above the horizon. It may also be determined by finding the meridian altitude of a point on the equator, which is the complement of the latitude.

7. **Correcting an Altitude — Refraction Correction.** — In measuring an altitude of any heavenly body it is necessary to apply a correction to the measured altitude to allow for the bending (refraction) of the rays of light in passing through the earth's atmosphere. This correction is always subtracted from the observed altitude to reduce it to the true altitude, because the ray of light is concave downward, and hence the object appears too high above the horizon. This correction in minutes of angle may be taken from Table I.

8. **Index Correction.** — In measuring altitudes with a transit the plate bubbles should not be relied upon, but after an altitude has been measured the vertical arc should be examined to see if the vernier reads 0° when the *telescope* bubble is in the middle of its tube. If it does not read 0° the reading of the vertical arc must be corrected by an amount equal to this error. If the 0 line of the vernier is on the same side of the 0° line of the arc at both readings, the index correction is to be subtracted; if the 0 line of the vernier passes the 0° graduation when the telescope is brought down to the horizontal position the correction is to be added.

* Published annually (three years in advance) by the Bureau of Equipment, Navy Department, Washington, D. C.

9. **Making Solar Observations.** — In making observations on the sun with the surveyor's transit it is desirable to have a dark glass placed over the eyepiece to cut down the light so that it will not be too bright for the eye. If no such dark glass accompanies the instrument the observation can be made by throwing the sun's image on a piece of paper held behind the eyepiece. If the objective is focussed on a distant object and the eyepiece tube drawn out, the image of the sun and the cross-hairs can be seen on the paper. The disc can be sharply focussed by moving the paper toward or away from the eyepiece. By this device observations can be made almost as well as by means of the dark glass.

10. **Making Star Observations.** — In making observations at night it is necessary to illuminate the field of view of the telescope in order to make the cross-hairs visible. This should be done in such a manner as to avoid heating either the instrument or the air just in front of the telescope. If the instrument is heated the adjustments will be disturbed; if the air is heated the image will appear very unsteady. With some instruments there is a special reflector placed in a shade tube, fitting to the objective slide, so that when a lantern is held at one side of the telescope the light is reflected into the tube. If no such reflector is at hand a satisfactory result may be obtained by placing a piece of tracing cloth, or oiled paper, in front of the objective, folding the edges back over the tube and fastening it in place by means of a rubber band. A hole about one-half inch in diameter should be cut in the centre of the paper so that light from the star can pass through the central portion of the lens while the outer edge of the lens is covered. The cross-hairs will then be made visible by light diffused by the tracing cloth.

For a few minutes just before dark Polaris can be easily seen through the telescope before it can be seen with the unaided eye, and the cross-hairs will be visible against the sky without artificial illumination. In order to find Polaris under these circumstances it is necessary to know its approximate altitude at the time. The telescope must be focussed on some very distant object and then raised until the vernier indicates the star's altitude. By pointing the telescope about north and then moving the instrument very slowly right and left the star can be found. For the method of finding the altitude of the star see Arts. 19 to 21, and equation [7], p. 29. Observations on Polaris just at dusk can be utilized when making observations by the method of Art. 26.

If accurate results are desired in the observations for azimuth the instrument should be firmly set up and allowed to stand for some time before the observations are begun; the observations should be made as quickly as is consistent with careful work, as delay simply allows the instrument more opportunity to change its position and thus introduce error. Unless the transit is in perfect adjustment it is well to make two observations, one with the telescope direct and the other with the telescope reversed, and to use the mean result.

11. **Azimuth Mark.** — In finding the azimuth of a transit line by star observations the instrument is set up at one end of the line, and at the other end is placed some sort of azimuth mark on which pointings can be made. If it is inconvenient or impossible to set this mark at the other end of the transit line it may be set in any convenient position, not too near the instrument, and its azimuth from the instrument determined. This observed direction is then connected with the survey by means of an angle measured in the daytime between the line to the azimuth mark and the transit line. In this case the mark should be so arranged that it can be sighted on either in the daytime or at night. For an accurate determination of the azimuth the mark usually consists of a box with a small hole cut in the side toward the observer so that light from a lantern placed inside can shine through the opening. The diameter of the hole should be such as to subtend an angle of about one second (0.3 inch per mile). The light should have the appearance of a small point of light like a star; it should not appear large or blurred. If the line to be sighted over is not one of the lines of the survey but is to be connected with the survey by an angle measured in the daytime, the box should have a target or a stripe painted on it to serve as a mark when sighting on it in the daytime. The centre of the hole and the centre of the target should coincide and should be placed carefully on the line to be sighted over. For the best results the mark should be placed far enough away so that the focus of the telescope does not have to be altered when changing from the star to the mark.

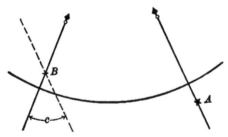

Fig. 1. Convergence of the Meridians.

12. **Convergence of Meridians.** — In comparing azimuth observations made at points in different longitudes it will be necessary to allow for the angular convergence of the meridians at the two places. In running

westward, in the northern hemisphere, the meridians are turned farther toward the right, as shown in Fig. 1.

If an azimuth observation were made at A and a traverse run westward to B and another azimuth observation made at point B, then it would be necessary to add to the azimuth observed at B the correction c in order to reduce this azimuth to what it would be if referred to the meridian at A. The difference between this corrected azimuth and the azimuth computed from the angles represents the accumulated error of the angular measurements of the survey. The amount of this correction for convergence of the meridians is shown in Table II.

METHODS OF OBSERVING.

13. **Azimuth from an Observed Altitude of the Sun.** — In the following paragraphs it is assumed that the instrument is set up at some regular transit station of a survey and that it can be sighted at another station. The lower motion of the transit should remain clamped during the observation. The telescope is pointed at the sun, and the sun's image "found" in the field and sharply focussed. When a dark glass is used the cross-hairs usually cannot be seen except when they appear against the sun's disc. The telescope should be moved up and down a little so that the cross-hairs and also the stadia hairs can be identified (against the sun's disc) in order to avoid observing on a wrong hair. The observation is made by measuring the altitude of the upper and lower edges of the sun's disc and measuring the horizontal angles to the right and left edges so that the mean of these pairs of observations gives the altitude of the sun's centre and a horizontal angle to the centre corresponding to the same instant. In the first half of the observation the vertical cross-hair is set tangent to the left edge of the sun (by means of the upper plate tangent screw), and the horizontal cross-hair is set tangent to the upper edge of the sun; the vertical arc and the horizontal circle are then read and recorded. The watch time of the observation is also noted. The accuracy of the result may be increased by taking several such pointings and using the mean. In the second half of the observation the vertical cross-hair is set tangent to the right edge and the horizontal cross-hair tangent to the lower edge, thus placing the sun in the opposite quarter of the field from that used before. Both angles and the time are again recorded. The same number of pointings should be made in the second half as in the first. The telescope should then be levelled and the vernier examined to see if there is any index correction to be applied to the readings of the vertical arc. If the plate is set so that the vernier reads azimuths, as is customary in stadia surveying, then these vernier readings

alone will give the horizontal angle between the sun and the meridian from which the azimuths are being read. If the circle is not set for azimuths the upper clamp should be loosened and the telescope sighted on some line of the survey and the vernier read again, so that this vernier reading, combined with the two readings on the sun, will give the horizontal angle between the sun and the station sighted.

Theoretically it is immaterial whether the observations are made in the exact order given above or not, provided the sun is observed first in one of the quadrants formed by the cross-hairs and then in the *opposite* quadrant. It will be found, however, that the sun moves so rapidly that it is difficult to set both cross-hairs accurately in position at the same instant, hence the observation will be easier and also more accurate if we select that pair of opposite quadrants in which it will be necessary to make but one setting with the tangent screw, the other setting being made by the motion of the sun itself. This may be done in the following manner. If the observation is to be made in the forenoon set the vertical cross-hair a little in advance of the left edge of the sun (see the lower

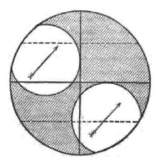 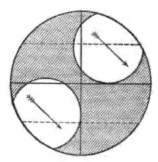

Fig. 2a.   Fig. 2b.
A.M. Observations.   *P.M. Observations.*

Diagram showing position of sun's disc a few seconds before the instant of observation for azimuth (northern hemisphere).

The arrows show the direction of the sun's motion. Stadia hairs are shown as dotted lines.

part of Fig. 2a), and keep the horizontal cross-hair tangent to the upper edge of the sun by means of the vertical tangent screw, following it until the left edge of the sun has moved up to the vertical cross-hair. At this instant stop moving the tangent screw, note the time, and read the angles. For the right and lower edges the position is as shown in the upper part of Fig. 2a, the horizontal cross-hair cutting across the lower portion of the sun and the vertical cross-hair being kept tangent to the right edge by means of the plate tangent screw. For observations in the afternoon the positions will be as shown in Fig. 2b. If a transit with an inverting eyepiece is being used these positions will of course appear reversed.

If the transit is provided with stadia hairs care should be taken not to mistake one of them for the middle cross-hair. In the figure the stadia hairs are represented by dotted lines, the distance between them being 1/100 part of the focal length of the objective. The angular distance between them is therefore about 0° 34′, a little greater than the diameter of the sun.

If the transit has a complete vertical circle the telescope should be reversed between the two pointings on the sun to eliminate errors of adjustment.

14. Computing the Azimuth from Sun Observations. — In order to compute the azimuth we proceed as follows:

1.) ALTITUDE. — Take the mean of the altitudes of the upper and lower edges of the sun, and subtract from it the refraction correction taken from Table I, thus obtaining the true altitude. The index correction must also be applied.

2.) HORIZONTAL ANGLE. — Take the mean of the vernier readings of the horizontal circle for the pointings on the sun.

If a pointing has been made on some reference mark, take the difference between this vernier reading and the mean vernier reading for the sun, the result being the horizontal angle between the mark and the sun.

3.) DECLINATION. — Take from the Nautical Almanac * (or any solar ephemeris) the *declination* † of the sun at *Greenwich Mean Noon* (G.M.N.) for the date of the observation, and also the *difference for 1 hour* and its algebraic sign, found in the next column to the right. In order to obtain the declination at the instant of the observation it will be necessary to allow for the change in the declination since the instant of Greenwich Mean Noon. If the watch keeps *Standard Time* the correction for change in declination can be made in a very simple manner. At the instant of Greenwich Mean Noon it is 7 A.M. Eastern Time, 6 A.M. Central Time, 5 A.M. Mountain Time, and 4 A.M. Pacific Time. Hence we can obtain the time elapsed since G.M.N. by simply subtracting 7 A.M., 6 A.M., etc., as the case may be, from the observed watch time, first adding

---

* It is not necessary to use the large Nautical Almanac for obtaining the sun's declination. Pamphlets containing the declination and the equation of time are issued by the Hydrographic Office (Publication No. 118) and may be obtained from the regular agents. Copies of the solar ephemeris are also published in the form of handbooks for engineers and for the use of navigators. Values of the sun's declination for the years 1909 to 1912 inclusive will be found in Table XV.

† This is given in the Almanac under the heading *Apparent Declination*.

12^h to the time if it is afternoon. The difference for 1 hour taken from the Almanac is to be multiplied by this elapsed time expressed in hours and the result added to or subtracted from the declination at G. M. N. (This multiplication may be avoided by the use of Table XIV.) An examination of the declination for the preceding or following dates will show whether it is increasing or decreasing and hence show whether the correction is to be added or subtracted. If the declination is considered positive when the sun is north of the equator and negative when south, then the elapsed time multiplied by the difference for 1 hour as given in the Almanac is always to be added. For example, suppose the declination is desired for Nov. 10, 1909, at 2^h 30^m P.M., Eastern time. The decl. for G. M. N., Nov. 10, $= - 17° 03'.1$, the diff. for 1 hour $= - 42''.5$. The time elapsed since G. M. N. $= 2^h$ 30^m P.M. $+ 12^h - 7$ A.M. $= 14^h$ $30^m - 7^h = 7^h.5$. The total change is $- 42''.5 \times 7^h.5 = - 5'.3$. The corrected declination is $- 17° 03'.1 - 5'.3 = - 17° 08'.4$. The hourly change never exceeds 1 minute of angle, so that if the watch is in error by as much as 10 minutes the resulting error in the declination will have a small effect on the azimuth.

If the watch keeps *local time* the watch time of G. M. N. is found by subtracting from 12^h the west longitude of the place expressed in hours, minutes, and seconds. For example, if the observer were in longitude 93° W. his (local) noon would occur 6^h 12^m after G. M. N., i.e., 5^h 48^m A.M. by his watch (if correct) is the instant of G. M. N.

4.) AZIMUTH. — Compute the azimuth of the sun, from the south point, by the formula

$$\log \text{vers Az.} = \log [\sin \{90° - (\text{Lat.} + \text{Alt.})\} + \sin \text{Decl.}]$$
$$+ \log \sec \text{Lat.}$$
$$+ \log \sec \text{Alt.}, \qquad [1]$$

in which Az. is the azimuth of the sun's centre from the south point, east or west; Lat. is the latitude of the place either taken from a map to the nearest minute or obtained by observation (see Art. 17); Alt. is the corrected altitude of the sun's centre; and Decl. is the declination of the sun at the instant of the observation. The arrangement of the computation is as shown in Examples 1 to 3, pp. 11 and 12. The latitude and the altitude are written down, their sum taken, and its complement written beneath. From Table III we take the log secants of the latitude and the altitude. These are found by looking up the secant for the next smaller angle in the left portion of the table and adding the proportional parts for the minutes from the proper column at the right. The log secant can thus be written down directly, to the nearest unit in the fourth figure.* The characteristics (o for all the log secants occurring within the limits of this table) have been omitted in Table III. From Table IV we obtain the natural sine of 90°—(latitude +altitude)† and also natural sin declination and take their algebraic sum. If the declination is -- the sine is —. From Table V we look up the log of this sum and add it to the two log secants. This sum is the log vers of the azimuth reckoned from the south point. In Table VI are given the log versed sines, the arrangement being exactly as in the preceding tables. If the observation is made in the afternoon the angle from Table VI is the azimuth desired; if in the forenoon the angle must be subtracted from 360 degrees, since azimuths are reckoned from the south point in a clockwise direction. This azimuth, combined with the measured horizontal angle, will give the azimuth of the line desired.

If it is desired to compute the azimuth with as great precision as the observations will afford, *i.e.*, to about 5 or 10 seconds, tables carried to five places should be used in the computation.

The formula given above applies to the northern hemisphere, but if the algebraic sign of the declination is changed and the azimuth reckoned from the north point instead of the south it will apply to the southern hemisphere. In the southern hemisphere the positions of the sun shown in Fig. 1 will obviously be changed.

15. **When to Observe.** — The most favorable times for accurate observations by this method are when the sun is nearly east or west. If the best results are desired the observations should *not* be made within 2 hours of noon nor when the sun's altitude is much less than 10 degrees.

* The table extends only to 60 degrees, which is sufficient for all ordinary cases. Should it be necessary to find the log secant of an angle greater than 60 degrees it may be done by taking the natural sine of the complement (Table IV), looking up its log (Table V) and subtracting this log from zero. For angles greater than about 80 degrees, however, this method is not sufficiently accurate.

† The sine is employed rather than cos (latitude + altitude) so that all numbers may be taken from the tables in exactly the same manner. If the sum of the latitude and the altitude exceeds 90 degrees the natural sine of this angle is negative. (See Example 2.)

EXAMPLE 1.

OBSERVATION ON THE SUN FOR AZIMUTH.

Latitude 42° 21′ N.   Nov. 28, 1905.

Instrument at Sta. 110.

Point sighted Station 111	Hor. Circle	Vert. Arc	Watch (Eastern Time)
	238° 14′		
⊙ (left)	311 48	14° 41′	8ʰ 41ᵐ A.M.
⊙ (left)	312 20	15 00	
⊙ (right)	312 27	15 55	
⊙ (right)	312 51	16 08	8 47
Mean	312° 21′.5	Obs. Alt. 15° 26′	8ʰ 44ᵐ
	238° 14		
Horizontal Angle	74° 07′.5	Refr. 3.5	7ʰ
		True Alt. 15° 22′.5	1ʰ 44ᵐ Gr. Time

COMPUTATION.

Lat. 42° 21′   log sec .1313
Alt. 15 22.5   log sec .0158

Sum 57° 43′.5

Sun's Decl. at G. M. N. = − 21° 14′.9
Diff. 1ʰ = − 26″.81

Nat. sin   .5340   Co 32° 16′.5
Nat. sin − .3626   Decl.− 21° 15′.7

26″.81 × 1ʰ.73 =   − 0′.8

Decl. at 8ʰ 44ᵐ = − 21° 15′.7

Alg. Sum .1714   log 9.2340

log vers 9.3811

Az. of Sun 40° 34′.7
Sta. 111 N. of Sun 74° 07′.5*

114° 42′.2
Az. 245° 18′
Sta. 110 to Sta. 111, N. 65° 18′ E.

* The plate readings (azimuths) indicate that Sta. 111 was to the left of the sun and hence North of it.

EXAMPLE 2.

AZIMUTH OBSERVATION ON SUN.

Latitude 42° 30′ N.   July 15, 1907.

Instrument at Sta. B.

	Hor. Circle	Vert. Arc	Eastern Time
Sta. 7	0° 00′		
⌊o	99° 19′	56° 35′	9ʰ 47ᵐ A.M.
⌊o	99° 54′	56° 54′	
o⌉	99° 40′	57° 40′	
o⌉	100° 07′	58° 02′	9 51
Mean	99° 45′	57° 20′	9ʰ 49ᵐ
		Refr. 0.5	
		57° 19′.5	

COMPUTATION.

Lat. 42° 30′   log sec .1324
Alt. 57 19.5 log sec .2677

sin — .1706   Co −9° 49′.5
sin   .3691   Decl. + 21° 39′.7

sum   .1985

Decl. at G. M. N. + 21° 40′.7
22″.8 × 2ʰ.8 =   − 1′.0

Decl. = + 21° 39′.7

log 9.2978

log vers 9.6979

Az. S. 59° 55′ E.
Sta. 7 north of sun 99 45

S. 159° 40′ E.
Sta. B to Sta. 7 = 200° 20′

The azimuth of Sta. B to Sta. 7 as calculated from the angles of the survey is 200° 18′.5

EXAMPLE 3.

AZIMUTH OBSERVATION AT ⊙ 25, Aug. 6, 1907, 5ʰ 04ᵐ P.M. in latitude 42° 29′.2 N.
Mean Alt. = 22° 29′.3. Mean plate reading = 92° 35′ (supposed to be true azimuth).

COMPUTATION.

Decl. at G. M. N. + 16° 57′.6
40″.7 × 10ʰ.1 =   − 6′.9

Decl.   + 16° 50′.7

Lat. 42° 29′.2   log sec .1323
Alt. 22° 29.3   log sec .0343

Sum 64° 58′.5

nat. sin .4230 Co − 25° 01′.5
nat. sin .2898 Decl. + 16° 50′.7

Sum .7128

log 9.8530

log vers 0.0196
Azimuth 92° 39′
Vernier 92 35

Error =   04′

Hence the azimuths read at ⊙ 25 are 4′ too small.

If the azimuth calculated by the preceding rule exceeds 70° the azimuth from the *north point* may be calculated as follows: Take the difference between the Latitude and the Altitude, and subtract it from 90°. From the natural sin of this angle subtract the natural sin Declination. The log of the result added to the log secants of the Latitude and Altitude gives the log vers Az. measured from the north. This affords a convenient means of checking azimuths between 70° and 110°. It will sometimes be found that the two azimuths will differ 1′, or even 2′, owing to the fact that only four places are used in the logarithms. The mean of the two results will always be more accurate than the result of a single computation.

16. **Azimuth by Altitude of a Star.** — The azimuth of a star can be determined in the same way as the azimuth of the sun, provided the star can be identified and its declination obtained. Since a star has no appreciable diameter its image should be bisected with both cross-hairs. Any of the brighter stars contained in the list given in the Nautical Almanac * can be used for this observation. For accurate results the star's declination should not be greater than about + 20 degrees nor less than about − 20 degrees, and at the time of the observation the star should be nearly due east or due west. Since the declination of a fixed star does not change appreciably in 24 hours it will not be necessary to note the time as in a solar observation.

17. **Observation for Latitude.** — In order to obtain the sun's azimuth it is generally necessary to know the latitude of the place within about 1 minute. In many cases this can be scaled from some reliable map with sufficient accuracy. If no such map is available the latitude must be observed directly, either by the sun's altitude at noon or by the altitude of the pole-star. If we can find the distance of the point of observation north or south of some other point whose latitude is known, the latitude of the instrument may be found by taking 6080 feet equal to 1 minute of latitude. The latitude can often be found in this way in places which have been surveyed by the United States Public Lands System, since the latitude of some of the points can be ascertained and the distance north or south to other points found from the township and section numbers.

18. **Latitude by the Sun at Noon.** — This observation is made by measuring the altitude of the sun at noon, when it is a maximum. The transit should be set up and levelled some time before

---

* For a condensed list see table of Fixed Stars in the Nautical Almanac under the heading "Mean Places"; for the exact places, right ascension and declination for any date, see table of "Apparent Places." A short list is given on p. 28 of this book.

noon* and the horizontal cross-hair set on the sun's lower edge. As
long as the sun continues to rise it should be followed with the vertical
motion of the telescope, keeping the cross-hair exactly tangent to the
lower edge of the disc. As soon as the sun begins to drop below the cross-
hair the motion of the tangent screw should be stopped and the vertical arc read.

This altitude must be corrected for (1) index error, (2) refraction,
(3) semi-diameter of the sun, and (4) the sun's declination. The refraction may be taken from Table I. The sun's semi-diameter may be taken
from the Nautical Almanac, but for approximate results may be taken
as 16′ in March and in September, 16′ 15″ in December, and 15′ 45″
in June. If the lower edge of the sun is observed the correction is
to be added to the measured altitude; the upper edge could have been
observed, in which case the semi-diameter should be subtracted. The
declination of the sun is found as described in Art. 14, or it may be
found by taking from the Nautical Almanac the declination for Greenwich
Apparent Noon, multiplying the difference for 1 hour by the number
of hours in the longitude and adding this to the declination at Greenwich
Apparent Noon. The declination must be subtracted from the altitude
if the sun is north of the equator (+), added if south (−). The altitude
thus found is the complement of the latitude.

EXAMPLE. OBSERVATION FOR LATITUDE BY ALTITUDE OF ☉ AT NOON, Jan. 13,
1905. Longitude 4ʰ 45ᵐ W. (approx.). Watch time = 11ʰ 53ᵐ (Eastern Time).

First Method.

Decl. at G. M. N. =	− 21° 32′.6
25″.2 × 4ʰ.9 =	+ 2 .1
Decl.	− 21° 30′.5

Second Method

Decl. at Apparent Noon =	− 21° 32′ 31″
25″.2 × 4.75 =	+ 2′ 00″
Decl. at Local Noon	− 21° 30′ 31″

Maximum Altitude ☉ 25° 55′
Refraction   2′
———
25° 53′
Semi-diameter   16 .3
———
Altitude of centre 26° 09′.3
Declination − 21 30 .5
———
Complement of Latitude 47° 30′.8
Latitude 42° 20′.2

* It should be remembered that the time of the sun's maximum altitude may differ
considerably from noon by the watch. To obtain the Standard Time of this observation,
call the time of the observation 12ʰ, Apparent Time; reduce this Apparent Time to Mean
Time by adding or subtracting the *equation of time* as given in the Nautical Almanac; then
reduce the Mean Time to Standard Time by taking the difference in longitude (expressed
in h. m. s.) between the place and the standard meridian and *adding* it if the place is *west*
of the standard meridian, *subtracting* if the place is *east*.

19. Azimuth and Latitude by Observation of ∂ Cassiopeiæ and Polaris. — The method described in the following articles is applicable when only approximate results are desired, say within about 1 minute of the true values, and when it is desired to obtain the result quickly and without using the Nautical Almanac. An advantage offered by this method is that it is not necessary to know the local time, since this is determined with sufficient accuracy by the observation itself. Tables VII to XII are so arranged that all of the quantities needed in this observation may be found by interpolation, and usually all the tables required for an observation appear at the same opening of the book.

20. Finding the Stars. — In order to make this observation it is necessary to be able to identify certain stars near the north pole. The most conspicuous constellation in the northern sky is the *Great Dipper*, or *Great Bear* (*Ursa Major*). (See Fig. 3.) Polaris, the star on which the azimuth observation chiefly depends, is readily found by reference to the Great Dipper. The two stars forming the side of the dipper bowl which is farthest from the handle are called the *pointers* because a line through them points very nearly to Polaris, the distance to Polaris being about five times the distance between the pointers. On the opposite side of the pole from the Great Dipper is the constellation *Cassiopeia*, shaped like a letter W. The star ∂ (delta) Cassiopeiæ, which is to be used in this observation, is the one at the bottom of the first stroke of the W, i.e., the lower left-hand star when the W is right side up. The *Little Dipper* is an inconspicuous constellation; Polaris is at the end of the dipper handle, and two fairly bright stars form the outer side of the dipper bowl. The other stars in this constellation are quite faint. Another star which will be referred to later is ∂ Draconis. When ∂ Cassiopeiæ is above the pole ∂ Draconis will be found west (left) of the meridian at about the same altitude as Polaris, these three stars forming a right triangle (nearly), the right-angle being at Polaris. The distance from Polaris to ∂ Draconis is less than the distance from Polaris to ∂ Cassiopeiæ. It will be observed that ∂ Draconis, Polaris, and the lower star in the bowl of the Little Dipper form a triangle which is nearly equilateral. ∂ Draconis is not as bright as ∂ Cassiopeiæ. There are several faint stars near ∂ Draconis, one of which might possibly be confused with it. This other star (ε Draconis) is nearly on a line drawn from ∂ Draconis to ∂ Cassiopeiæ, and its distance from ∂ Draconis is about 4 degrees, a little less than the distance between the *pointers*. ε Draconis is not as bright as ∂ Draconis.

21. Explanation of Method. — In order to determine the azimuth of Polaris and the latitude of the place it is necessary to find by some means the position of Polaris with respect to the pole at the instant of the observation. This depends upon the hour angle of Polaris at the time of the

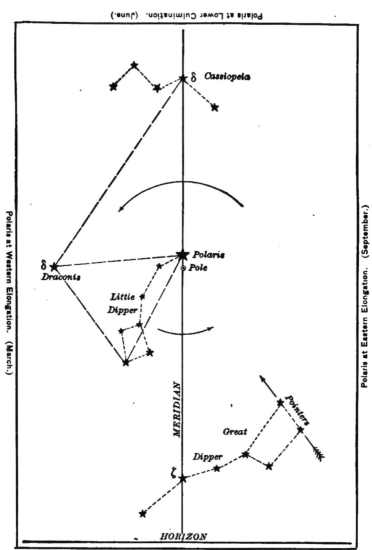

DIAGRAM SHOWING THE CONSTELLATIONS ABOUT THE NORTH POLE.

The arrows show the direction of the apparent motion of the stars.

FIG. 3.

observation. In order to determine the coördinates of Polaris, i.e., its distance above or below the pole and its distance east or west of the meridian, we measure the altitude of the star δ Cassiopeiæ and also the altitude of Polaris. From these altitudes we can calculate the coördinates of Polaris. By referring to Fig. 4 it will be seen that the pole, Polaris, and δ Cassiopeiæ are all nearly in the same plane (i.e., on the same hour circle), the two stars being on the *same* side of the pole. The direction of the apparent motion of the stars is shown by the arrow. Hence the relative position of δ Cassiopeiæ and Polaris as seen by the observer is at once a key to the position of the pole itself. If δ Cassiopeiæ is directly above Polaris, then Polaris is above the pole and nearly in the meridian; if δ Cassiopeiæ is below and to the left of Polaris, the latter is below and to the left of the pole. We may think of these two stars, then, as moving around the pole together as though they were two points on an arm pivoted at the pole.

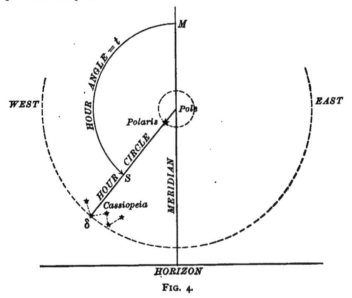

FIG. 4.

If we know the latitude of the observer, and the polar distance and the altitude of δ Cassiopeiæ at any instant, we may calculate the hour angle of this star, i.e., the arc MS in Fig. 4. Hence if Polaris were exactly on the same hour circle with δ Cassiopeiæ the hour angle of Polaris would be the same as that computed for δ Cassiopeiæ. In reality Polaris has a slightly smaller hour angle than δ Cassiopeiæ, the difference between the two increasing slowly from year to year. This interval is 6^m 58^s for the year 1910, 10^m 57^s for 1920, and 15^m 13^s for 1930. After we observe the altitude of δ Cassiopeiæ we may wait until this interval of time has elapsed and then make the observation on Polaris, the latter *then* being in the position it would have occupied at the first observation if the two stars were on the same hour circle. This instant at which Polaris is to be observed we may call for convenience the *computed time*. When the hour angle of Polaris is known for the instant of the observation the

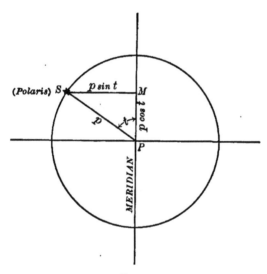

FIG. 5. Coördinates of Polaris.

coördinates may be found at once. In Fig. 5 if p is the polar distance of Polaris and t its hour angle, then

$$PM = p \cos t$$

and
$$SM = p \sin t. \tag{2}$$

PM is the amount (nearly) which Polaris is above or below the pole, hence

$$\text{Lat.}* = \text{Alt. of pole} = \text{Alt. of Polaris} - p \cos t. \tag{3}$$

* For a more accurate expression for the latitude we should add to the above series the quantity K, from Table XIII. See equation [7].

If cos t is given its proper algebraic sign in the different quadrants this equation holds true for all positions of the star. The quantities PM and SM may be found in Tables VIII and XI.

The coördinate $p \sin t$ is the angular distance of the star east or west of the meridian. The azimuth of Polaris depends not only upon the distance $p \sin t$ but also upon the altitude of the star as seen by the observer; it may be found by the equation

$$\text{Azimuth} = p \sin t \sec \text{ altitude,} \qquad [4]$$

or $\qquad \text{Azimuth} = p \sin t + p \sin t \text{ exsec altitude.*} \qquad [5]$

The azimuth is computed by taking from Tables IX or XII the azimuth correction ($p \sin t$ exsec altitude) and adding it to $p \sin t$. This azimuth is, of course, reckoned from the north point.

When δ Cassiopeiæ is near the meridian, i.e., nearly above or nearly below Polaris, an accurate determination of its hour angle cannot be made. In this case another star, δ Draconis (p. 15), can be substituted for δ Cassiopeiæ and the hour angle of Polaris derived in a similar manner, so that it is almost always possible to use this method.

22. **The Tables.** — All of the quantities needed in this observation may be taken from Tables VII to XII. In Table VII are given the hour angles of δ Cassiopeiæ for different altitudes and different latitudes. The hour angle for any latitude and any altitude at which an accurate observation can be made may be found by interpolation in this table. If the star is east of the meridian the tabular hour angle should be subtracted from 360 degrees to obtain the true hour angle. Since it is necessary to interpolate in this table both for the altitude and the latitude it will be simpler and also more accurate to observe the star when the altitude is a *whole degree*, preferably an *even numbered degree*, and thus confine the interpolation to the latitudes. If the observation on Polaris follows the observation on δ Cassiopeiæ by the interval of time corresponding to the date, as given above, then the hour angle of Polaris at the time it was observed is the same as the hour angle taken from Table VII. In Table VIII will be found the values of $p \sin t$ and $p \cos t$ for different hour angles of Polaris between 30 degrees and 150 degrees, and for the years 1910, 1920, and 1930. If the date falls between those given in the table it will be necessary to interpolate to obtain the coördinates for the date of the observation. With the value of $p \sin t$ found in Table VIII and the measured altitude, we take from Table IX the correction $p \sin t$ exsec altitude. This correction, added to $p \sin t$, gives the azimuth desired.

* The external secant, or, the secant minus unity.

It will be noticed that in general we do not know the latitude of the place, and therefore we cannot determine the hour angle in a direct manner as was assumed above. If the latitude is not known we may proceed as follows. From the relative position of δ Cassiopeiæ and Polaris we estimate the latitude, remembering that Polaris is $1°$ $10'$ from the pole, and its direction from the pole is nearly the same as the direction of δ Cassiopeiæ from Polaris. With this approximate latitude and the measured altitude we take from Table VII an approximate hour angle of δ Cassiopeiæ. With this approximate hour angle we take from Table VIII the value of $p \cos t$. This is the correction to be applied to the altitude to give the corrected latitude. (Equation [3].) With this new latitude we take from the table a more accurate value of the hour angle. Using this new hour angle we find from Table VIII the value of $p \sin t$. Two approximations will always give $p \sin t$ with sufficient accuracy. The azimuth correction from Table IX may be taken out as before. If the latitude as well as the azimuth is desired the above process of approximation may be continued if necessary until the value of $p \cos t$ agrees within about 1 minute with the preceding value. The latitude is then found by equation [3].

It is not really necessary to observe Polaris at the same hour angle as δ Cassiopeiæ, although this simplifies the calculation slightly. In case the observation is not made at the *computed time* we must correct the hour angle accordingly before taking out $p \cos t$ from Table VIII. The correction is made by converting this difference in the time interval into degrees and adding it to or subtracting it from the hour angle of δ Cassiopeiæ. Suppose for instance that the observation on Polaris was made 5^m 20^s before the calculated time, then 5^m $20^s = 5^m.33 = 1°.33$ (see Art. 5), which must be subtracted from the hour angle of δ Cassiopeiæ to obtain the hour angle of Polaris, since Polaris had not reached this hour angle at the instant it was observed.

23. **Making the Observations.** — Set up the instrument at one end of the line whose azimuth is to be found, and set a lantern or arrange an azimuth mark at the other end of the line. See if δ Cassiopeiæ is in a favorable position for an observation. If δ Cassiopeiæ is near the meridian, either above or below the pole, an accurate observation cannot be made on this star. If the altitude of the star and the latitude of the observer are such that an hour angle can be found in Table VII, then a reliable observation can be made. If it is found that δ Cassiopeiæ can be used, point the telescope at this star, examine the vertical circle to see what the approximate altitude is, and set it so that the vernier reads a whole degree (preferably an even numbered degree) + the refraction correction for this altitude (Table I). If the star is west of the meridian

it is moving downward, and the telescope must be set at some altitude below that of the star. If the star is east of the meridian the telescope must bo set at a higher altitude than that of the star. Watch the star and note the time when it crosses the horizontal cross-hair. The star moves so slowly that the observation is not precise, but it is sufficiently exact for this purpose. Next *calculate the time* of the observation to be made on Polaris by adding to the watch time just noted the interval from the top of Table VII. Set the plate vernier at o degrees and point at the azimuth mark, using the lower clamp. Loosen the upper clamp, point the telescope toward Polaris, and set both cross-hairs on the star. Follow the star's motion, using the vertical tangent screw and the upper plate tangent screw, until the *computed time* is indicated by the watch, then see that both cross-hairs are bisecting the star. Read the vertical arc and the plate vernier, and determine the index correction to the vertical angle. The latitude and the azimuth are then found from Tables VII to IX as described on pp. 19 and 20.

The method of using these tables is illustrated by the following examples. The first two illustrate observations in which Polaris was observed at the calculated instant. In Examples 3 and 4 Polaris was observed before the computed time arrived.

EXAMPLE 1. OBSERVATION ON POLARIS AND δ CASSIOPEIÆ FOR LATITUDE AND AZIMUTH.

March 14, 1908.

Set telescope at altitude 26° 02′ (26° + the refraction correction for 26°); δ Cassiopeiæ passed horizontal cross-hair at $9^h 06^m 10^s$. Interval for 1908 = $6^m 10^s$ (Table VII, top). Computed time for observation on Polaris = $9^h 12^m 20^s$. Set on mark with vernier at 0° 00′. Bisected Polaris with both cross-hairs at $9^h 12^m 20^s$. Altitude = 41° 52′.5. Index correction = − 0′.5. Horizontal angle, 38° 31′.5. Mark is west of Polaris. δ Cassiopeiæ is below and west of Polaris.

FIRST APPROXIMATION.

Approx. Lat. = 42°   Alt. 41° 52′.5
    True Alt. = 26°   Index corr.   − 0.5
                            ――――――
From Table VII, hour angle, $t = 112°.0$ 41° 52′.0
                            Refr.   − 1.1
                            ――――――
                            True Alt. 41° 50′.9
From Table VIII, for 112°, we find $p \cos t = − 26′.6$ $p \cos t$ − 26.6
                            ――――――
                            Approx. Lat. 42° 17′.5

SECOND APPROXIMATION.

From Table VII, using lat. 42° 17′.5, $t = 112°.7$
From Table VIII, for 112°.7, $p \cos t = − 27′.3$
and $p \sin t = 65′.6$ Alt. 41° 50′.9
    From Table IX, az. corr. = 22 .5 $p \cos t$ − 27.3
                            ――――――
        Az. = 88′.1 = 1° 28′.1   Lat. 42° 18′.2
    Measured angle = 38° 31′.5

Azimuth of mark is   N. 39° 59′.6 W.

EXAMPLE 2. OBSERVATION ON POLARIS AND δ CASSIOPEIÆ. August 18, 1908.

Set telescope at altitude 34° 01′.5 (34° + refraction corr.). δ Cassiopeiæ passed horizontal cross-hair at 9ʰ 07ᵐ 11ˢ. Interval for 1908 = 6ᵐ 10ˢ. Calculated time for observation on Polaris = 9ʰ 13ᵐ 21ˢ. Set on mark with vernier at 0° 00′. Set on Polaris at 9ʰ 13ᵐ 21ˢ. Altitude = 41° 46′. Index correction = 0. Horizontal angle = 89° 38′. Mark east of Polaris. δ Cassiopeiæ below and east of Polaris.

Approx. Lat. 42°			Alt. =	41° 46′
True Alt. 34°			Refr. =	1′.1
Table VII, hour angle =	92°.9			41° 44′.9
True hour angle = 360° − 92°.9 =	267°.1		$p \cos t$ =	− 3.6
			Approx. Lat. =	41° 48′.5

From Table VIII, for 92°.9, $p \cos t$ = − 3′.6

Table VII, lat. 41° 48′.5, t =	92°.6
Table VIII, for 92°.6, $p \cos t$ = −	3′.2
$p \sin t$ = 71′ =	1° 11′
Table IX, az. corr. =	24.2

True az. =	1° 35′.2		
Horizontal angle =	89 38		41° 44′.9
			− 3′.2
Angle from North	91° 13′.2	Lat. =	41° 48′.1
Bearing of mark	S. 88° 46′.8 E.		

EXAMPLE 3. OBSERVATION ON POLARIS AND δ CASSIOPEIÆ. Feb. 11, 1908.

Set telescope at altitude 55° 01′ (55° + refraction corr.). δ Cassiopeiæ passed horizontal cross-hair at 7ʰ 03ᵐ 05ˢ. Interval for 1908 = 6ᵐ 10ˢ. Computed time of observation on Polaris = 7ʰ 09ᵐ 15ˢ. At 7ʰ 07ᵐ 05ˢ Polaris was bisected with both cross-hairs. Altitude = 43° 05′. Index correction = +1′. Angle between Polaris and mark = 67° 11′. Mark is east of Polaris. δ Cassiopeiæ is above and west of Polaris.

Estimated latitude = 42°	
True altitude = 55°	
From Table VII, t = 49°.9	
Interval = 0.5	
Hour angle of Polaris = 49°.4	
From Table VIII, $p \cos t$ = +45′.6	Computed time 7 − 09 − 15
	Observed time 7 − 07 − 05
∴ Latitude = 42° 19′.4	Interval = 2ᵐ 10ˢ
	= 0°.5

SECOND APPROXIMATION.

Corrected hour angle =	50°.4
Interval =	.5
Hour angle of Polaris =	49°.9
Table VIII, $p \sin t$ =	54′.4
Table IX, corr. =	19.9
Azimuth of Polaris =	74′.3
=	1° 14′
Measured angle = 67° 11′	

Direction of mark = N. 65° 57′ E.

EXAMPLE 4. OBSERVATION ON POLARIS AND δ CASSIOPEIÆ. Jan. 2, 1908.

δ Cassiopeiæ — Alt. 59° 11'.5; time, 9ʰ 07ᵐ 50ˢ. Interval, 6ᵐ 10ˢ. Computed time, 9ʰ 14ᵐ 00ˢ. Polaris — Alt. 43ᶜ 13'; time, 9ʰ 11ᵐ 00ˢ. Angle from mark to Polaris, 106° 49'; mark west of star. δ Cassiopeiæ is above and W. of Polaris.

Assumed latitude = 42°	Computed time 9ʰ 14ᵐ 00ˢ
True altitude = 59° 11'	Observed time 9 11 00
Table VII, t = 41°.4	
Table VIII, p cos t = +53'.4	Interval = 3ᵐ 00ˢ
∴ Latitude = 42° 19' (approx.)	

SECOND APPROXIMATION.

Table VII, corrected t = 41°.5	
Interval, 3ᵐ = .8	
Hour angle of Polaris = 40°.7	Alt. 43° 13'
Table VIII, p sin t = 46'.3	Refr. 1'
Table IX, corr. = 17.2	43° 12'
63'.5	p cos t +53.3
Az. of Polaris = 1° 03'.5	Lat. 42° 18'.7
Horizontal angle = 106° 49'	

N. 107° 52'.5 W.
Az. of mark S. 72° 07'.5 W.

24. Observations on δ Draconis. — When δ Cassiopeiæ has an hour angle of less than 30 degrees or more than 150 degrees the hour angle cannot be accurately determined from the measured altitude, and consequently Table VII does not include such hour angles. In this case the altitude of δ Draconis may be observed, since this star is nearly always in a favorable position for an observation at times when δ Cassiopeiæ is in an unfavorable position. The observation is made in just the same way as for δ Cassiopeiæ except for the time interval between the two observations. The difference in hour angle of δ Draconis and Polaris is 6ʰ 14ᵐ 21ˢ for 1910 — too long to wait — hence we make the observation on Polaris as soon as convenient after the altitude of δ Draconis has been measured, and correct the hour angle of δ Draconis as previously described for δ Cassiopeiæ. For example, if the altitude of δ Draconis was taken at 8ʰ 15ᵐ P.M. in the year 1910, adding 6ʰ 14ᵐ 21ˢ gives 14ʰ 29ᵐ 21ˢ as the calculated time when Polaris will have the same hour angle as δ Draconis. If the observation were made on Polaris at 8ʰ 20ᵐ 00ˢ P.M. we must subtract from the true hour angle of δ Draconis the quantity 14ʰ 29ᵐ 21ˢ — 8ʰ 20ᵐ 00ˢ = 6ʰ 09ᵐ 21ˢ = 92°.3, which will give the hour angle of Polaris at the time it was observed. The difference in hour angle between Polaris and δ Draconis for different dates is as follows: — 1910, 6ʰ 14ᵐ 21ˢ = 93°.6; 1920, 6ʰ 18ᵐ 58ˢ = 94°.7; 1930, 6ʰ 23ᵐ 54ˢ = 96°.0. The following examples will illustrate the method of making the calculations.

Example 5.

Jan. 10, 1908.

Set telescope at Altitude 37° 01'. δ Draconis passed horizontal hair at 5ʰ 57ᵐ 10ˢ. Interval for 1908 = 6ʰ 13ᵐ 26ˢ. Observed Polaris at 5ʰ 59ᵐ 40ˢ; altitude = 43° 30'; plate vernier, 275° 27'.5; mark, 216° 13'.0. δ Draconis west of Polaris.

Estimated latitude = 42°	Observed time 5ʰ 57ᵐ 10ˢ
True altitude = 37°	Interval 6 13 26
Table X, Hour Angle = 93°.3	Computed time 12 10 36
Hour Angle of Polaris* = 93°.3 − 92°.7 = +0°.6	Observed time 5 59 40
Table XI, $p \cos t = +1°$ 11'	Diff. 6ʰ 10ᵐ 56ˢ
∴ Latitude = 42° 18'	= 92°.7
Table X, Corrected Hour Angle = 94°.1	
Correction for interval = 92°.7	Vernier Readings.
Hour Angle of Polaris = 1°.4	275° 27'.5
	216 13
	Angle = 59° 14'.5

Table XI, $p \sin t = $ 01'.7		Alt. 43° 30'	
Table XII, correction =	.6	Refr.	1'
Azimuth Polaris =	02'.3		43° 29'
Angle to mark =	59° 14'.5	$p \cos t$	+1 11
Azimuth of mark = N. 59° 16'.8 W.		Lat.	42° 18'

Example 6.

Jan. 2, 1908.

δ Draconis; — Observed altitude = 27° 25'; observed time = 8ʰ 32ᵐ 45ˢ. Interval = 6ʰ 13ᵐ 26ˢ; computed time = 14ʰ 46ᵐ 11ˢ; Polaris; — altitude = 43° 20'; observed time = 8ʰ 34ᵐ 45ˢ. Angle, mark to Polaris, 107° 01'. Mark is west of Polaris. δ Draconis is west of Polaris.

Assumed latitude = 42°	Computed time = 14ʰ 46ᵐ 11ˢ
True altitude = 27° 23'	Observed time = 8 34 45
Table X, t = 124°.0	Interval = 6 11 26
Hour angle of Polaris = 124°.0 − 92°.8 = 31°.2	= 92°.8
Table VIII for 31°.2, $p \cos t$ = +60'	
∴ Latitude = 42° 19' (approx.)	
Table X, corrected t = 125°.2	
Hour angle of Polaris = 32°.4	Corrected altitude = 43° 19
	$p \cos t$ = +1° 00'

Table VIII, $p \sin t$ =	38'.0	Latitude = 42° 19'
Table IX, correction =	14 .2	
Azimuth of Polaris =	52'.2	
Angle =	107° 01'	
	107° 53'.2	
Azimuth of mark = S. 72° 07' W.		

* In case this Hour Angle becomes negative it should be subtracted from 360 degrees. Polaris in that case would be east of the meridian.

The above examples have been worked out more elaborately than would be required in many cases. Frequently the latitude will be known or may be estimated closely, so that only one approximation is needed. Nearly all of the interpolation may be done mentally.

25. **Meridian Line by Polaris at Culmination.** — When Polaris is near the meridian (i.e. near *culmination*) it will sometimes be convenient to use the following simple method, which will give the meridian with about the same accuracy as the method of Articles 19–24. We may use in this case either δ Cassiopeiæ or the star ζ in the Great Dipper (see Fig. 3). If we determine by means of a surveyor's transit the instant when one of these stars, say δ Cassiopeiæ, is vertically above or vertically below Polaris, then we have only to wait a certain interval of time, depending upon the date, when Polaris will be *in the meridian*. If Polaris is sighted at this instant, the telescope may be lowered and the direction of this line marked on the ground, thus giving a meridian line without

CORRECTION.

Page 25, Table A, first line, last column, for 5.5^{m} read 6.5^{m}

may be neglected. A convenient way to keep this interval small is to set on Polaris, lower the telescope, and wait until the other star appears in the field; then reset on Polaris, lower the telescope, and observe the instant of transit. The intervals which it is necessary to wait before Polaris is in the meridian are given in the following table for the two stars men-

TABLE A.

Date.	Interval between δ Cassiopeiæ and Polaris.	Interval between ζ Ursæ Majoris and Polaris.
1910	$7^{\mathrm{m}}.3$	$5^{\mathrm{m}}.5$
1920	11 .5	10 .3

tioned and for the years 1910 and 1920. These intervals are nearly correct when Polaris is either above or below the pole. In high latitudes it will in general be necessary to use the star which is at lower culmination at the time of the observation. In low latitudes it may be more convenient to use the star which is at upper culmination.

26. **Azimuth by Polaris when the Time is Known.** — If the error of the watch and the longitude of the place are known within about 1^m the azimuth of Polaris at any time may be found as follows. If the time is P.M. use the method described under (a); if A.M. use method (b).

(a) 1. If the time is P.M. and the watch keeps Standard Time, convert the observed watch time into Greenwich Mean Time by adding 5^h, 6^h, etc., according to whether the watch keeps Eastern Time, Central Time, etc.

(a) 2. Increase this time by 10^s for each hour in the interval and add to the result the "Right Ascension of the Mean Sun" at Greenwich Mean Noon of the same date, thus obtaining Greenwich Sidereal Time.

(b) 1. If the time is A.M. first add 12^h and then add 5^h, 6^h, etc., as before, to obtain the Greenwich Mean Time.

(b) 2. Add 10^s per hour as before, and then add to this the Right Ascension of the Mean Sun for the *preceding* date, thus obtaining Greenwich Sidereal Time.

3. Convert the Greenwich Sidereal Time (found by (a) or (b)) into Local Sidereal Time by subtracting* from it the longitude of the place expressed in hours, minutes, and seconds.

4. From this Local Sidereal Time subtract the Right Ascension of Polaris† for the date. The result is the hour angle of Polaris for the instant considered.

5. Take from the Almanac the Declination† of Polaris for the date, and subtract it from 90 degrees, obtaining the polar distance p. Then compute the azimuth by the equation

$$Az. = p \sin t \sec h. \qquad [4]$$

If the hour angle is less than 12^h the star is west of the meridian; if greater than 12^h it is east of the meridian.

---

* If a large angle is to be subtracted from a smaller angle the latter may be increased by 24^h in order to make the subtraction possible.

† Found in the table of Circumpolar Stars in the Nautical Almanac.

EXAMPLE.

Feb. 11, 1908.

At $7^h 07^m 05^s$, angle between Polaris and mark is $67° 11'$; Altitude Polaris $= 43° 06$. Mark is east of Polaris. Longitude $= 71° 04\frac{1}{2}'$ W. $= 4^h 44^m 18^s$ W. Watch keeps Eastern Time.

Watch	$7^h 07^m 05^s$ P.M.
	5
Greenwich Time	$12^h 07^m 05^s$ (after noon)
Correction	$2^m 01^s$
Right Ascension Mean Sun	21 20 38
Greenwich Sidereal Time*	$33^h 29^m 44^s$
	24
Greenwich Sidereal Time	$9^h 29^m 44^s$
Longitude	4 44 18
Local Sidereal Time	$4^h 45^m 26^s$
Right Ascension Polaris	1 25 32
Hour Angle of Polaris	$3^h 19^m 54^s$
	$t = 49° 59'$

Declination $= +88° 49' 09''$
$p = 1° 10' 51''$
$= 70'.85$

$\log p = 1.8503$
$\log \sin t = 9.8842$
$\log \sec h = 0.1365$

$\log$ Azimuth $= 1.8710$
Azimuth of Polaris $= 74'.31$
$= 1° 14' 19''$ W.

Angle $67° 11'$
Azimuth of Polaris $1° 14' 19'$
Azimuth of mark N. $65° 56' 41''$ E.

27. Accurate Determination of Azimuth by Observation on Polaris. — An accurate determination of azimuth may be made by using a method similar to that of Article 19 except that the determination of the hour angle must be more precise, and the angle between the pole-star and the azimuth mark must be measured with greater accuracy.

The determination of the hour angle should be made by observing several altitudes in quick succession, with their corresponding watch readings, on some star which can be identified (called the *time-star*) and which is nearly east or west at the time of the observation. Following is a list of bright stars which may be used for time determinations in the northern hemisphere. The position of these stars may be found by consulting the star maps on pp. 62–3. The exact right ascension and declination for any date must of course be taken from the list of Apparent Places given in the Nautical Almanac. In identifying stars by means of the chart the observer should be on the lookout for the planets. These are not fixed in position and hence cannot be shown on the chart. If a very bright star is seen which cannot be found on the chart it is a planet and should not be used for time observations unless it can be positively identified and its exact position for the date determined.

---

* If the Greenwich Sidereal Time exceeds 24^h it should be decreased by 24^h. If the Greenwich Sidereal Time is less than the longitude it must be increased by 24^h to make the subtraction possible.

LIST OF FIXED STARS

Constellations and Letters.	Name.	Right Ascension.	Declination.
α Arietis.........	Hamal..........	2^h.01^m	+23° 02′
α Tauri.........	Aldebaran.......	4 30	+16 19
β Orionis........	Rigel...........	5 10	− 8 18
α Orionis........	Betelgeux........	5 50	+ 7 23
α Can. Maj.......	Sirius..........	6 41	−16 35
α Can. Min.......	Procyon.........	7 34	+ 5 27
α Hydræ.........	Alphard.........	9 23	− 8 15
α Leonis........	Regulus.........	10 03	+12 25
β Leonis........	Denebola........	11 44	+15 05
α Virginis.......	Spica..........	13 20	−10 41
α Boötis.........	Arcturus.......	14 11	+19 40
α Cor. Bor.......	Alphecca........	15 30	+27 01
α Ophiuchi.......	Ras-Alhague......	17 30	+12 38
α Lyræ..........	Vega...........	18 33	+38 42
α Aquilæ........	Altair..........	19 46	+ 8 57
α Cygni.........	Arided.........	20 38	+44 57
α Pegasi.........	Markab.........	23 00	+14 42

28. Determining the Hour Angle. — If we know the latitude of the place, the declination of the star,* and the mean of the altitudes (corrected for refraction) we can compute the hour angle of the star by the formula

$$\log \text{vers } t = \log [\sin \{90° - (\text{Lat.} - \text{Decl.})\} - \sin \text{Alt.}]$$
$$+ \log \sec \text{Lat.}$$
$$+ \log \sec \text{Decl.} \quad [6]$$

This is of the same general form as equation [1] and may be solved by means of Tables III to VI. The value of t thus found is the hour angle of the time-star corresponding to the mean of the observed times.

In order to find the hour angle of Polaris we take from the Nautical Almanac the right ascension of Polaris and the right ascension* of the time-star for the date of the observation; the difference between these is the same as the difference between the hour angles of the two stars at any time. Combining this difference of hour angle with the computed hour angle of the time-star we obtain the hour angle of Polaris at the instant T, the mean of the watch readings. If the right ascension of the time-star is less than that of Polaris the difference is to be subtracted from the hour angle of the time-star to obtain the hour angle of Polaris. If the right ascension of the time-star is greater than the right ascension of Polaris the difference is to be added to the hour angle of the time-star to obtain the hour angle of Polaris.

* Found in the table of Fixed Stars, Apparent Places, Nautical Almanac.

Since errors of the vertical angle will produce errors in the computed hour angle it is advisable to make observations on two time-stars, one of which is east of the meridian and one west. The mean of the two results for the hour angle of Polaris will be nearly free from such errors.

29. Determining the Azimuth. — We now measure, by several repetitions, the angle between Polaris and the azimuth mark, noting the time at each pointing on the star. For the best results a set of repetitions should be made with the telescope direct and another set with the telescope reversed. Several such sets may be made, if desired, to increase the accuracy. The altitude of Polaris should be measured before and after each set.

30. Computing the Azimuth. — In computing the azimuth of the mark from the transit it will be better to reduce each half-set separately. In the first half-set take the mean of the observed watch times (on Polaris) and call this T', which corresponds (nearly) to the average angle between the star and the mark in this half-set. The interval $T' - T$ is the amount which the hour angle of Polaris has increased since the instant T, when the time observation was made. Strictly speaking this interval is in solar units of time and should be reduced to sidereal units. This reduction may be made with sufficient accuracy by increasing the interval $T' - T$ by 1ˢ for each 6ᵐ 05ˢ of time in the interval, i.e., if 18ᵐ 15ˢ elapsed between the two observations the true interval is 18ᵐ 15ˢ + 3ˢ = 18ᵐ 18ˢ. This corrected interval added to the hour angle of Polaris at the time T gives the hour angle of Polaris at the time T', when the angles were measured. The azimuth of the star at this instant T' may be found with sufficient accuracy by the formula

$$\text{Azimuth} = p \sin t \sec h, \qquad [4]$$

in which

p is the polar distance for the date of the observation, taken from the Nautical Almanac.

t is the hour angle of Polaris, just computed, at the time T', and

h is the altitude of the star at the time T', found by interpolating between the altitudes measured just before and just after the angle measurements. The azimuth is $+$ if the star is west of the meridian, $-$ if east, and is reckoned from the north point of the horizon. If the latitude of the place is known and it is not convenient to measure the altitude when the observations are made, it may be found by the equation

$$\text{Altitude} = \text{Latitude} + p \cos t - K, \qquad [7]$$

where $p \cos t$ is the quantity given in Tables VIII and XI, and K is a correction found in Table XIII. The second half-set of angles is to be reduced in a similar manner. The azimuth of the star for each half-set is to be combined with the mean angle obtained from the repetitions, thus giving the azimuth of the mark. The interval of time during a half-set should be kept as short as possible if the best results are desired, because the azimuth of the star at the mean of the observed times is not strictly the same as the mean of the different azimuths, on account of the curvature of the star's path.

The determination of the hour angle may be made by several other methods, such as observing the transit of the time-star across the meridian,* across the vertical circle through Polaris, or by equal altitudes of two stars.

The following examples will serve to illustrate the method of computing the azimuth:

EXAMPLE 1.

AZIMUTH OBSERVATION.

Approximate latitude 42° 21'.   Feb. 11, 1908.

OBSERVATION FOR THE TIME.

Altitudes of *Regulus* (east).	Times.
17° 05'	7ʰ 12ᵐ 16ˢ
17 31	14 31
17 49	16 07
18 02	17 20
Mean 17° 37'	$T = 7^h 15^m 03\frac{1}{2}^s$
Refraction 3'	
Alt. 17° 34'	

HORIZONTAL ANGLES FROM AZIMUTH MARK TO POLARIS.

(Mark east of star.)

Tel. Direct		Times—on Polaris	Altitude
Mark	0° 00'	7ʰ 20ᵐ 38ˢ	43° 03'
3d rep. 201° 48'		23 00	
		23 56	
Mean	67° 16'	$T' = 7^h 22^m 31^s$	
Tel. Reversed		7ʰ 27ᵐ 09ˢ	
Mark	0° 00'	28 17	
3d rep. 201° 54'		29 21	43° 01'
Mean	67° 18'	$T' = 7^h 28^m 16^s$	

* An approximate meridian, found by one of the methods previously given, will usually be sufficiently accurate for this time observation.

COMPUTATION OF THE HOUR ANGLE OF REGULUS.

	Latitude 42° 21′	log sec.	.1313
	Declination +12° 25′	log sec.	.0102

	Difference 29° 56′	
nat. sin = .8666	Co 60° 04′	
nat. sin = .3018	Altitude 17° 34′	
Diff. = .5648		log 9.7519
		log vers = 9.8934

Right Ascension Regulus = 10ʰ 03ᵐ 29ˢ.1
Right Ascension Polaris = 1 25 32.4

Diff. Right Ascension = 8ʰ 37ᵐ 56ˢ.7
                      = 129° 29′

$t = -- 77° 26′$ at 7ʰ 15ᵐ 03ˢ
   $= 129° 29′$

Hour Angle of Polaris } $= 52° 03′$
 at 7ʰ 15ᵐ 03ˢ

HOUR ANGLES OF POLARIS

Intervals			
1st half-set	2nd half-set	Hour Angle of Polaris	Corresponding Time
$T' = 7ʰ 22ᵐ 31ˢ$	7ʰ 28ᵐ 16ˢ	53° 55′	7ʰ 22ᵐ 31
$T = 7$ 15 03	7 15 03	55° 22′	7 28 16
$T' - T =$ 7ᵐ 28ˢ	13ᵐ 13ˢ		
Red. to sidereal } 1	2		
Interval = 7ᵐ 29ˢ	13ᵐ 15ˢ		
= 1° 52′	3° 19′		
52° 03′	52° 03′		
Hour Angle 53° 55′	55° 22′		

COMPUTATION OF AZIMUTH.

First half-set.	Second half-set.
$p = 70′.85$	
log $p = 1.8503$	1.8503
log sin $t = 9.9075$	9.9153
log sec $h = .1361$	.1359
log azimuth = 1.8939	1.9015
azimuth = 78′.32	70′.71
azimuth = 1° 18′.32	1° 19.71
Angle = 67° 16′	67° 18′.00
Azimuth of line. N. 65° 57′.7 E.	N. 65° 58′.3 E.

Mean azimuth of mark = 245° 58′.0

Example 2.

Azimuth Observation.

Latitude 42° 03′   Sept. 5, 1906.

Horizontal Angles from Mark to Polaris.

(Mark East of Star)

Tel. Direct	Times — on Polaris.
Vernier	6ʰ 39ᵐ 48ˢ
0° 00′ 00″	40 57
6th Rep. 211° 45′ 30″	41 39
Mean 35° 17′ 35″	42 29
	43 13
	44 01

Mean 6ʰ 42ᵐ 01ˢ.1 = T'

Tel. Reversed	
0° 00′ 00″	6ʰ 50ᵐ 15ˢ
6th Rep. 211° 32′ 00″	51 16
Mean 35° 15′ 20″	52 52
	54 26
	55 37
	56 55

Mean 6ʰ 53ᵐ 33ˢ.5 = T'

T = 7ʰ 52ᵐ 26ˢ (mean)

Time Observations.

Mean Altitude of Arcturus = 27° 12′
Refraction =   2′
Reduced Altitude = 27° 10′

Computation of Hour Angle of Arcturus.

Latitude = 42° 03′   log sec. .1292
Declination = 19 40.4   log sec. .0263

Difference = 22° 22′.6

nat. sin = .9247   Co = 67° 37′.4
nat. sin = .4566   Altitude = 27° 10′

Diff. = .4681   log   9.6703

log vers = 9.8258

Right Ascension Arcturus = 14ʰ 11ᵐ 22ˢ.6 t = 70° 42′ at 7ʰ 52ᵐ 26ˢ
Right Ascension Polaris = 1 26 21.7   191 15

Diff. Right Ascension = 12ʰ 45ᵐ 00ˢ.9   Hour Angle } = 261° 57′ at 7ʰ 52ᵐ 26ˢ
= 191° 15′   of Polaris }

HOUR ANGLES OF POLARIS.

Intervals

	1st half-set	2nd half-set	Hour Angle of Polaris	Time
T'	$6^h\ 42^m\ 01^s.1$	$6^h\ 53^m\ 33^s.5$		
T	$7\ \ 52\ \ 26$	$7\ \ 52\ \ 26$	$244°\ 18'$	$6^h\ 42^m\ 01^s$
$T' - T =$	$-1^h\ 10^m\ 25^s$	$-0^h\ 58^m\ 52^s$	$247°\ 11'$	$6^h\ 53^m\ 33^s.5$
Red. to Sidereal	12	10		
Interval =	$-1^h\ 10^m\ 37^s$	$-0^h\ 59^m\ 02^s$		Altitudes
=	$-\ 17°\ 39'$	$-\ 14°\ 46'$	Latitude = $42°\ 03'.0$	$42°\ 03'.0$
	$261°\ 57'$	$261°\ 57'$	$p\cos t = \ -31\ .1$	-27.8
	$244°\ 18'$	$247°\ 11'$	$41°\ 31'.9$	$41°\ 35'.2$
			K (Table XIII) $.5$	$.5$
			Altitude $41°\ 31'.4$	$41°\ 34'.7$

COMPUTATION OF THE AZIMUTH.

$$p = 1°\ 11'\ 47''.1 = 71'.78$$

	Direct	Reversed
$p \cos t =$	$-31'.13$	$-27'.84$
$\log p \cos t =$	1.4932 n.	1.4446 n.
$\log \cos t =$	9.6372 n.	9.5886 n.
$\log p =$	1.8560*	1.8560
$\log \sin t =$	9.9548 n.	9.9646 n.
$\log \sec h =$	0.1257	0.1261
$\log$ Azimuth =	1.9365 n.	1.9467 n.
Azimuth = $-$	$86'.40$	$-88'.45$
=	$-\ 1°\ 26'\ 24''$	$-\ 1°\ 28'\ 27''$
Angle	$35\ \ 17\ \ 35$	$35\ \ 15\ \ 20$
Azimuth of line =	$36°\ 43'\ 59''$	$36°\ 43'\ 47''$

Mean, N 36° 43′ 53″.0 E.

31. Meridian by Polaris at Elongation. — If Polaris is at its extreme east or west *elongation*, i.e., if it has its greatest east or west bearing, its azimuth can be accurately determined without knowing the exact time at which elongation occurs. The approximate time, near enough to determine when the observation should be begun, may be taken from Table B. The positions of the constellations at the times of elongations may be seen by reference to Fig. 3.

* $p \cos t$ is computed by adding upward, and the azimuth by adding downward.

TABLE B.*

APPROXIMATE TIMES OF ELONGATION OF POLARIS COMPUTED
FOR THE 90th MERIDIAN WEST OF GREENWICH, FOR THE
YEAR 1907.

Date.	Eastern Elongation.		Western Elongation.	
1907	h	m	h	m
Jan. 1	0	49	12	39
" 15.................................	23	50	11	44
Feb. 1	22	42	10	36
" 15.................................	21	47	9	41
Mar. 1	20	52	8	46
" 15.................................	19	57	7	51
Apr. 1	18	50	6	44
" 15.................................	17	55	5	40
May 1	16	52	4	46
" 15.................................	15	57	3	51
Jun. 1	14	50	2	44
" 15.................................	13	56	1	50
Jul. 1	12	53	0	47
" 15.................................	11	58	23	48
Aug. 1	10	52	22	42
" 15.................................	9	57	21	47
Sept. 1	8	50	20	40
" 15.................................	7	55	19	45
Oct. 1	6	52	18	42
" 15.................................	5	58	17	48
Nov. 1	4	51	16	41
" 15.................................	3	56	15	46
Dec. 1	2	52	14	42
" 15.................................	1	57	13	47

In order to make this observation set the transit in position a half hour
or more before the pole-star reaches its eastern or western elongation.
Set the vertical cross-hair on Polaris and follow it, with the plate tangent
screw, as long as the star continues to move away from the meridian.
When the star is near its greatest elongation it will appear to move vertically for a few minutes and then will begin to move back toward the
meridian. While the star is in this extreme position it should be carefully

* To find the time for any other date than the first or 15th interpolate between the
values given in the table, the daily change being about 4 minutes. The table may be
used to obtain the approximate times of elongation for other years. If the time given
is more than 12ʰ, subtract 12ʰ and call it A.M.; otherwise it is P.M.

bisected with the vertical cross-hair and a stake set at some convenient distance from the instrument in line with the cross-hair. In order to eliminate the errors due to poor adjustment of the transit the telescope should be immediately reversed, the star again bisected and another point set in line with the vertical cross-hair. The mean of these two points will give the direction of the star at elongation. The angle between this direction and the direction of the meridian may be calculated by the formula

$$\text{sin azimuth} = \text{sin polar distance} \times \text{sec latitude}, \qquad [8]$$

or, with sufficient accuracy,

$$\text{azimuth (in seconds)} = \text{polar distance (in seconds)} \times \text{sec latitude}. \quad [9]$$

This calculated azimuth may be laid off either by means of the transit, using repetitions, or by measuring the distance from the transit to the point that was set, and then calculating the perpendicular offset which will give a point exactly north of the transit. This offset should be laid off with a steel tape. If desired, angles to some mark may be measured instead of setting stakes in line with the cross-hair. At elongation Polaris changes its azimuth so slowly that there will usually be time to set several points or to measure several angles before the star changes its bearing as much as 5 seconds of angle.

32. **Meridian by Equal Altitudes of a Star.** — A very simple method of determining the direction of the meridian is by observing on a star at equal altitudes on opposite sides of the meridian. This method is accurate and requires no Nautical Almanac or tables of any kind; it is not convenient, however, as it requires two observations at night separated by several hours' time, but it may prove of value when for some reason the more rapid and convenient methods are not available, as, for instance, in the southern hemisphere where there is no bright star near the pole.

In order to determine the true meridian select some star which is not far from the pole and which is on the west side of the meridian in about the position of A, Fig. 6 The hour angle should be such that the star will reach an equal altitude on the east side of the meridian about 6h or 8h later, so that the second half of the observation will occur before daylight. One of the stars in Cassiopeia could be used, for example, the first observation being made when the star has an hour angle of about 135 degrees and the second at an hour angle of 225 degrees, as shown at A and A' in Fig. 6. The star is bisected with both cross-hairs, the horizontal circle is clamped, and the altitude is read and recorded. The

telescope is then lowered and a point set in line. Some memorandum
or sketch should be made for identifying the star at the second observation. When the star is approaching the same altitude on the east side of
the meridian the telescope is set at *exactly the same altitude* as was read at
the first observation. The star is then bisected with the vertical cross-
hair, and followed until it passes the horizontal cross-hair. After this
instant the tangent screw should not be touched. Another point is then
set in line with the vertical cross-hair as before. The bisector of the
angle between these two points is the meridian line through the instru-

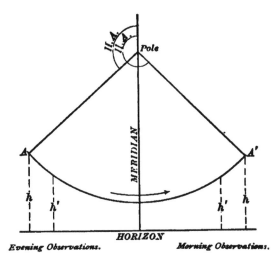

FIG. 6. Meridian by Equal Altitudes of a Circumpolar (Northern Hemisphere)

ment. If desired several pairs of altitudes may be observed, to increase
the accuracy, as h and h, h' and h', Fig. 6. Each pair should be combined
independently of the others. If preferred, angles may be measured
from some reference mark instead of setting points in line with the star.
Instead of taking the altitudes at random it is well to set the telescope so
that the vernier reads some whole degree or half degree and then set the
vertical cross-hair on the stár and follow it with the upper tangent screw
until both cross-hairs bisect the star. If the transit has a full vertical
circle, errors in the instrument may be eliminated by taking the evening
observations with the instrument direct and the morning observations
with the instrument reversed.

33. Meridian by Equal Altitudes of the Sun. —The observation just described may also be made on the sun at equal altitudes in the forenoon and afternoon, the difference being that the sun changes its declination during the interval between the observations and hence a correction must be applied in order to obtain the direction of the true meridian. Since the change for a given date is practically the same each year no Nautical Almanac is necessary, but the hourly change in declination for any date may be taken from Table XV.

The observation is made as follows: At some time in the forenoon, say between 8^h and 10^h A.M., set up the transit and set the plate vernier at o degrees. Point the telescope at some azimuth mark and clamp the lower plate. Loosen the upper plate and turn to the sun. Set the vertical arc so that the vernier reads some whole degree or some even 10 minutes (higher than the sun); then set the vertical cross-hair on the left edge of the sun and follow, with the upper tangent screw, until the lower edge of the sun is in contact with the horizontal cross-hair. Note the time and read the altitude and the horizontal angle. In the afternoon, a little before the sun reaches this same altitude, set the vernier on o degrees and point again on the mark. Set the telescope so that the vernier of the vertical arc reads the *same* altitude as was used for the A.M. observation. When the sun comes into the field of the telescope set the vertical cross-hair on the right edge of the sun and follow it, with the upper plate tangent screw, until the lower edge of the sun is again in contact with the horizontal cross-hair. Note the time and read the horizontal angle.

The mean of the two readings of the horizontal angle is approximately the angle between the mark and the south point. Since, however, the sun's declination is not the same at the two observations it will be necessary to apply the correction

$$C = \frac{d}{2 \cos \text{Lat.} \times \sin \text{Hour Angle}}, \qquad [10]$$

in which d is the hourly change from Table XV multiplied by the number of hours between the observations, and the Hour Angle equals half this number of hours turned into degrees. In the table, the + sign indicates that the sun is going north, the − sign indicates that it is going south. If the sun is going north the mean of the two angles gives a point west of the true south; if the sun is going south the mean angle is to a point east of south.

EXAMPLE.

Latitude 42° 18′ N.   April 19, 1906.

A.M. Observation   **P.M. Observation**

Reading on mark 0° 00′   Reading on mark 0° 00′
Pointings on left and upper edges   Pointings on right and upper edges
Altitude 24° 58′   Altitude 24° 58′
Horizontal Angle* 357° 14′ 15″   Horizontal Angle* 162° 28′ 00″
Time 7ʰ 19ᵐ 30ˢ A.M.   Time 4ʰ 12ᵐ 15ˢ P.M.

½ elapsed time = 4ʰ 26ᵐ 22ˢ   Change in declination in
            t = 66° 35′ 30″   4ʰ 26ᵐ 22ˢ = +52″ × 4ʰ.44
                                                    = 230″.9
log sin t = 9.9627   Mean Plate Reading 259° 51′ 08″
log cos L = 9.8690   180°
          ————   ————
            9.8317   Uncorrected bearing 79° 51′ 08″
log 230″.9 = 2.3634   Correction   − 5′ 40″
          ————   ————
            2.5317   Corrected bearing S. 79° 45′ 28″ E.
Correction = − 340″.2 = − 5′ 40″.2   Azimuth = 280° 14′ 32″

* Read "clockwise" in each case.

TABLES

I. REFRACTION CORRECTION.

II. CONVERGENCE OF MERIDIANS.

III. LOGARITHMIC SECANTS.

IV. NATURAL SINES.

V. LOGARITHMS OF NUMBERS.

VI. LOGARITHMIC VERSED SINES.

VII. HOUR ANGLES OF δ CASSIOPELÆ.

VIII. COÖRDINATES OF POLARIS.

IX. CORRECTION FOR AZIMUTH.

X. HOUR ANGLES OF δ DRACONIS.

XI. COÖRDINATES OF POLARIS.

XII. CORRECTION FOR AZIMUTH.

XIII. VALUES OF K FOR COMPUTING THE ALTITUDE.

XIV. CORRECTION TO SUN'S DECLINATION.

XV. SUN'S DECLINATION FOR 1909, 1910, 1911 AND 1912.

TABLE I.   REFRACTION CORRECTION.

(In Minutes.)

(True Alt. = Meas'd Alt. − Refr.)

Alt.	Refr.	Alt.	Refr.	Alt.	Refr.
5°	9'.9	13°	4'.1	25°	2'.1
6	8.5	14	3.8	30	1.7
7	7.4	15	3.6	35	1.4
8	6.6	16	3.3	40	1.2
9	5.9	17	3.1	45	1.0
10	5.3	18	3.0	50	0.8
11	4.9	19	2.8	55	0.7
12	4.5	20	2.6	60	0.6

TABLE II.   CONVERGENCE OF THE MERIDIANS.

(In Minutes.)

Lat.	Miles (east or west).									
	1	2	3	4	5	6	.7	8	9	10
30°	0'.5	1'.0	1'.5	2'.0	2'.5	3'.0	3'.5	4'.0	4'.5	5'.0
35	0.6	1.2	1.8	2.4	3.0	3.6	4.2	4.9	5.5	6.1
40	0.7	1.5	2.2	2.9	3.6	4.4	5.1	5.8	6.5	7.3
45	0.9	1.7	2.6	3.5	4.3	5.2	6.1	6.9	7.8	8.7
50	1.0	2.1	3.1	4.1	5.2	6.2	7.2	8.3	9.3	10.3

TABLE III.—LOGARITHMIC SECANTS.

Deg.	Minutes						Proportional Parts								
	0′	10′	20′	30′	40′	50′	1′	2′	3′	4′	5′	6′	7′	8′	9′
0	.0000	.0000	.0000	.0000	.0000	.0000	0	0	0	0	0	0	0	0	0
1	.0001	.0001	.0001	.0001	.0002	.0002	0	0	0	0	0	0	0	0	0
2	.0003	.0003	.0004	.0004	.0005	.0005	0	0	0	0	0	0	0	0	1
3	.0006	.0007	.0007	.0008	.0009	.0010	0	0	0	0	0	0	1	1	1
4	.0011	.0011	.0012	.0013	.0014	.0015	0	0	0	0	1	1	1	1	1
5	.0017	.0018	.0019	.0020	.0021	.0023	0	0	0	0	1	1	1	1	1
6	.0024	.0025	.0027	.0028	.0029	.0031	0	0	0	1	1	1	1	1	1
7	.0032	.0034	.0036	.0037	.0039	.0041	0	0	0	1	1	1	1	1	1
8	.0042	.0044	.0046	.0048	.0050	.0052	0	0	1	1	1	1	1	2	2
9	.0054	.0056	.0058	.0060	.0062	.0064	0	0	1	1	1	1	1	2	2
10	.0066	.0069	.0071	.0073	.0076	.0078	0	0	1	1	1	1	2	2	2
11	.0081	.0083	.0086	.0088	.0091	.0093	0	1	1	1	1	2	2	2	2
12	.0096	.0099	.0101	.0104	.0107	.0110	0	1	1	1	1	2	2	2	3
13	.0113	.0116	.0119	.0122	.0125	.0128	0	1	1	1	2	2	2	3	3
14	.0131	.0134	.0137	.0141	.0144	.0147	0	1	1	1	2	2	2	3	3
15	.0151	.0154	.0157	.0161	.0164	.0168	0	1	1	1	2	2	2	3	3
16	.0172	.0175	.0179	.0183	.0186	.0190	0	1	1	1	2	2	3	3	3
17	.0194	.0198	.0202	.0206	.0210	.0214	0	1	1	2	2	2	3	3	4
18	.0218	.0222	.0226	.0230	.0235	.0239	0	1	1	2	2	3	3	3	4
19	.0243	.0248	.0252	.0257	.0261	.0266	0	1	1	2	2	3	3	4	4
20	.0270	.0275	.0279	.0284	.0289	.0294	0	1	1	2	2	3	3	4	4
21	.0298	.0303	.0308	.0313	.0318	.0323	1	1	2	2	3	3	4	4	5
22	.0328	.0333	.0339	.0344	.0349	.0354	1	1	2	2	3	3	4	4	5
23	.0360	.0365	.0371	.0376	.0382	.0387	1	1	2	2	3	3	4	4	5
24	.0393	.0398	.0404	.0410	.0416	.0421	1	1	2	2	3	3	4	5	5
25	.0427	.0433	.0439	.0445	.0451	.0457	1	1	2	2	3	4	4	5	5
26	.0463	.0470	.0476	.0482	.0488	.0495	1	1	2	3	3	4	4	5	6
27	.0501	.0508	.0514	.0521	.0527	.0534	1	1	2	3	3	4	4	5	6
28	.0541	.0547	.0554	.0561	.0568	.0575	1	1	2	3	3	4	5	5	6
29	.0582	.0589	.0596	.0603	.0610	.0617	1	1	2	3	4	4	5	6	6
30	.0625	.0632	.0639	.0647	.0654	.0662	1	2	2	3	4	5	5	6	7
31	.0669	.0677	.0685	.0692	.0700	.0708	1	2	2	3	4	5	5	6	7
32	.0716	.0724	.0732	.0740	.0748	.0756	1	2	2	3	4	5	6	6	7
33	.0764	.0772	.0781	.0789	.0797	.0806	1	2	2	3	4	5	6	7	7
34	.0814	.0823	.0831	.0840	.0849	.0858	1	2	3	3	4	5	6	7	8
35	.0866	.0875	.0884	.0893	.0902	.0911	1	2	3	4	5	5	6	7	8
36	.0920	.0930	.0939	.0948	.0958	.0967	1	2	3	4	5	6	7	7	8
37	.0977	.0986	.0996	.1005	.1015	.1025	1	2	3	4	5	6	7	8	9
38	.1035	.1045	.1055	.1065	.1075	.1085	1	2	3	4	5	6	7	8	9
39	.1095	.1105	.1116	.1126	.1136	.1147	1	2	3	4	5	6	7	8	9
40	.1157	.1168	.1179	.1190	.1200	.1211	1	2	3	4	5	6	7	9	10
41	.1222	.1233	.1244	.1255	.1267	.1278	1	2	3	4	6	7	8	9	10
42	.1289	.1301	.1312	.1324	.1335	.1347	1	2	3	5	6	7	8	9	10
43	.1359	.1371	.1382	.1394	.1406	.1418	1	2	4	5	6	7	8	10	11
44	.1431	.1443	.1455	.1468	.1480	.1493	1	2	4	5	6	7	9	10	11
45	.1505	.1518	.1531	.1543	.1556	.1569	1	3	4	5	6	8	9	10	12
46	.1582	.1595	.1609	.1622	.1635	.1649	1	3	4	5	7	8	9	11	12
47	.1662	.1676	.1689	.1703	.1717	.1731	1	3	4	6	7	8	10	11	12
48	.1745	.1759	.1773	.1787	.1802	.1816	1	3	4	6	7	9	10	11	13
49	.1831	.1845	.1860	.1875	.1889	.1904	1	3	4	6	7	9	10	12	13
50	.1919	.1934	.1950	.1965	.1980	.1996	2	3	5	6	8	9	11	12	14
51	.2011	.2027	.2043	.2059	.2074	.2090	2	3	5	6	8	10	11	13	14
52	.2107	.2123	.2139	.2156	.2172	.2189	2	3	5	7	8	10	12	13	15
53	.2205	.2222	.2239	.2256	.2273	.2290	2	3	5	7	9	10	12	14	15
54	.2308	.2325	.2343	.2360	.2378	.2396	2	4	5	7	9	11	12	14	16
55	.2414	.2432	.2450	.2469	.2487	.2506	2	4	6	7	9	11	13	15	17
56	.2524	.2543	.2562	.2581	.2600	.2620	2	4	6	8	10	11	13	15	17
57	.2639	.2658	.2678	.2698	.2718	.2738	2	4	6	8	10	12	14	16	18
58	.2758	.2778	.2799	.2819	.2840	.2861	2	4	6	8	10	12	14	16	19
59	.2882	.2903	.2924	.2945	.2967	.2988	2	4	6	9	11	13	15	17	19
60	.3010	.3032	.3054	.3077	.3099	.3122	2	4	7	9	11	13	16	18	20

TABLE IV. — NATURAL SINES.

Deg.	0'	10'	20'	30'	40'	50'	1'	2'	3'	4'	5'	6'	7'	8'	9'
			Minutes.						Proportional parts.						
0	.0000	.0029	.0058	.0087	.0116	.0145	3	6	9	12	15	17	20	23	26
1	.0175	.0204	.0233	.0262	.0291	.0320	3	6	9	12	15	17	20	23	26
2	.0349	.0378	.0407	.0436	.0465	.0494	3	6	9	12	15	17	20	23	26
3	.0523	.0552	.0581	.0610	.0640	.0669	3	6	9	12	15	17	20	23	26
4	.0698	.0727	.0756	.0785	.0814	.0843	3	6	9	12	15	17	20	23	26
5	.0872	.0901	.0929	.0958	.0987	.1016	3	6	9	12	15	17	20	23	26
6	.1045	.1074	.1103	.1132	.1161	.1190	3	6	9	12	14	17	20	23	26
7	.1219	.1248	.1276	.1305	.1334	.1363	3	6	9	12	14	17	20	23	26
8	.1392	.1421	.1449	.1478	.1507	.1536	4	6	9	11	14	17	20	23	26
9	.1564	.1593	.1622	.1650	.1679	.1708	3	6	9	11	14	17	20	23	26
10	.1736	.1765	.1794	.1822	.1851	.1880	3	6	9	11	14	17	20	23	26
11	.1908	.1937	.1965	.1994	.2022	.2051	3	6	9	11	14	17	20	23	26
12	.2079	.2108	.2136	.2164	.2193	.2221	3	6	9	11	14	17	20	23	26
13	.2250	.2278	.2306	.2334	.2363	.2391	3	6	8	11	14	17	20	23	25
14	.2419	.2447	.2476	.2504	.2532	.2560	3	6	8	11	14	17	20	22	25
15	.2588	.2616	.2644	.2672	.2700	.2728	3	6	8	11	14	17	20	22	25
16	.2756	.2784	.2812	.2840	.2868	.2896	3	6	8	11	14	17	20	22	25
17	.2924	.2952	.2979	.3007	.3035	.3062	3	6	8	11	14	17	20	22	25
18	.3090	.3118	.3145	.3173	.3201	.3228	3	6	8	11	14	17	20	22	25
19	.3256	.3283	.3311	.3338	.3366	.3393	3	5	8	11	14	16	19	22	25
20	.3420	.3448	.3475	.3502	.3529	.3557	3	5	8	11	14	16	19	22	24
21	.3584	.3611	.3638	.3665	.3692	.3719	3	5	8	11	14	16	19	22	24
22	.3746	.3773	.3800	.3827	.3854	.3881	3	5	8	11	13	16	19	21	24
23	.3907	.3934	.3961	.3987	.4014	.4041	3	5	8	11	13	16	19	21	24
24	.4067	.4094	.4120	.4147	.4173	.4200	3	5	8	11	13	16	18	21	24
25	.4226	.4253	.4279	.4305	.4331	.4358	3	5	8	10	13	16	18	21	24
26	.4384	.4410	.4436	.4462	.4488	.4514	3	5	8	10	13	16	18	21	23
27	.4540	.4566	.4592	.4617	.4643	.4669	3	5	8	10	13	15	18	21	23
28	.4695	.4720	.4746	.4772	.4797	.4823	3	5	8	10	13	15	18	20	23
29	.4848	.4874	.4899	.4924	.4950	.4975	3	5	8	10	13	15	18	20	23
30	.5000	.5025	.5050	.5075	.5100	.5125	3	5	8	10	13	15	18	20	23
31	.5150	.5175	.5200	.5225	.5250	.5275	2	5	7	10	12	15	17	20	22
32	.5299	.5324	.5348	.5373	.5398	.5422	2	5	7	10	12	15	17	20	22
33	.5446	.5471	.5495	.5519	.5544	.5568	2	5	7	10	12	15	17	19	22
34	.5592	.5616	.5640	.5664	.5688	.5712	2	5	7	10	12	14	17	19	22
35	.5736	.5760	.5783	.5807	.5831	.5854	2	5	7	9	12	14	17	19	21
36	.5878	.5901	.5925	.5948	.5972	.5995	2	5	7	9	12	14	16	19	21
37	.6018	.6041	.6065	.6088	.6111	.6134	2	5	7	9	12	14	16	18	21
38	.6157	.6180	.6202	.6225	.6248	.6271	2	5	7	9	11	14	16	18	20
39	.6293	.6316	.6338	.6361	.6383	.6406	2	4	7	9	11	13	16	18	20
40	.6428	.6450	.6472	.6494	.6517	.6539	2	4	7	9	11	13	15	18	20
41	.6561	.6583	.6604	.6626	.6648	.6670	2	4	7	9	11	13	15	17	20
42	.6691	.6713	.6734	.6756	.6777	.6799	2	4	6	9	11	13	15	17	19
43	.6820	.6841	.6862	.6884	.6905	.6926	2	4	6	8	10	13	15	17	19
44	.6947	.6967	.6988	.7009	.7030	.7050	2	4	6	8	10	12	14	17	19

Deg.	0'	10'	20'	30'	40'	50'	1'	2'	3'	4'	5'	6'	7'	8'	9'
			Minutes.						Proportional parts.						

TABLE IV. — (Continued).

Deg.	Minutes.						Proportional parts.								
	0'	10'	20'	30'	40'	50'	1'	2'	3'	4'	5'	6'	7'	8'	9'
45	.7071	.7092	.7112	.7133	.7153	.7173	2	4	6	8	10	12	14	16	18
46	.7193	.7214	.7234	.7254	.7274	.7294	2	4	6	8	10	12	14	16	18
47	.7314	.7333	.7353	.7373	.7392	.7412	2	4	6	8	10	12	14	16	18
48	.7431	.7451	.7470	.7490	.7509	.7528	2	4	6	8	10	12	14	15	17
49	.7547	.7566	.7585	.7604	.7623	.7642	2	4	6	8	9	11	13	15	17
50	.7660	.7679	.7698	.7716	.7735	.7753	2	4	6	7	9	11	13	15	17
51	.7771	.7790	.7808	.7826	.7844	.7862	2	4	5	7	9	11	13	14	16
52	.7880	.7898	.7916	.7934	.7951	.7969	2	4	5	7	9	11	12	14	16
53	.7986	.8004	.8021	.8039	.8056	.8073	2	3	5	7	9	10	12	14	15
54	.8090	.8107	.8124	.8141	.8158	.8175	2	3	5	7	8	10	12	14	15
55	.8192	.8208	.8225	.8241	.8258	.8274	2	3	5	7	8	10	12	13	15
56	.8290	.8307	.8323	.8339	.8355	.8371	2	3	5	6	8	10	11	13	14
57	.8387	.8403	.8418	.8434	.8450	.8465	2	3	5	6	8	9	11	12	14
58	.8480	.8496	.8511	.8526	.8542	.8557	2	3	5	6	8	9	11	12	14
59	.8572	.8587	.8601	.8616	.8631	.8646	1	3	4	6	7	9	10	12	13
60	.8660	.8675	.8689	.8704	.8718	.8732	1	3	4	6	7	9	10	12	13
61	.8746	.8760	.8774	.8788	.8802	.8816	1	3	4	6	7	8	10	11	13
62	.8829	.8843	.8857	.8870	.8884	.8897	1	3	4	5	7	8	9	11	12
63	.8910	.8923	.8936	.8949	.8962	.8975	1	3	4	5	6	8	9	10	12
64	.8988	.9001	.9013	.9026	.9038	.9051	1	3	4	5	6	8	9	10	11
65	.9063	.9075	.9088	.9100	.9112	.9124	1	2	4	5	6	7	8	10	11
66	.9135	.9147	.9159	.9171	.9182	.9194	1	2	4	5	6	7	8	9	10
67	.9205	.9216	.9228	.9239	.9250	.9261	1	2	3	4	6	7	8	9	10
68	.9272	.9283	.9293	.9304	.9315	.9325	1	2	3	4	5	6	8	9	10
69	.9336	.9346	.9356	.9367	.9377	.9387	1	2	3	4	5	6	7	8	9
70	.9397	.9407	.9417	.9426	.9436	.9446	1	2	3	4	5	6	7	8	9
71	.9455	.9465	.9474	.9483	.9492	.9502	1	2	3	4	5	6	7	7	8
72	.9511	.9520	.9528	.9537	.9546	.9555	1	2	3	4	4	5	6	7	8
73	.9563	.9572	.9580	.9588	.9596	.9605	1	2	2	3	4	5	6	7	7
74	.9613	.9621	.9628	.9636	.9644	.9652	1	2	2	3	4	5	5	6	7
75	.9659	.9667	.9674	.9681	.9689	.9696	1	1	2	3	4	4	5	6	7
76	.9703	.9710	.9717	.9724	.9730	.9737	1	1	2	3	3	4	5	5	6
77	.9744	.9750	.9757	.9763	.9769	.9775	1	1	2	3	3	4	4	5	6
78	.9781	.9787	.9793	.9799	.9805	.9811	1	1	2	2	3	3	4	5	5
79	.9816	.9822	.9827	.9833	.9838	.9843	1	1	2	2	3	3	4	4	5
80	.9848	.9853	.9858	.9863	.9868	.9872	0	1	1	2	2	3	3	4	4
81	.9877	.9881	.9886	.9890	.9894	.9899	0	1	1	2	2	3	3	3	4
82	.9903	.9907	.9911	.9914	.9918	.9922	0	1	1	2	2	2	3	3	3
83	.9925	.9929	.9932	.9936	.9939	.9942	0	1	1	2	2	2	2	3	3
84	.9945	.9948	.9951	.9954	.9957	.9959	0	1	1	1	1	2	2	2	3
85	.9962	.9964	.9967	.9969	.9971	.9974	0	0	1	1	1	1	2	2	2
86	.9976	.9978	.9980	.9981	.9983	.9985	0	0	1	1	1	1	1	1	2
87	.9986	.9988	.9989	.9990	.9992	.9993	0	0	0	1	1	1	1	1	1
88	.9994	.9995	.9996	.9997	.9997	.9998	0	0	0	0	0	0	0	1	1
89	.9998	.9999	.9999	.0000	.0000	.0000	0	0	0	0	0	0	0	0	0
Deg.	0'	10'	20'	30'	40'	50'	1'	2'	3'	4'	5'	6'	7'	8'	9'

Minutes.   Proportional parts.

TABLE V.— LOGARITHMS OF NUMBERS.

Nat. Nos.	0	1	2	3	4	5	6	7	8	9	Proportional parts								
											1	2	3	4	5	6	7	8	9
10	.0000	.0043	.0086	.0128	.0170	.0212	.0253	.0294	.0334	.0374	4	8	12	17	21	25	29	33	37
11	.0414	.0453	.0492	.0531	.0569	.0607	.0645	.0682	.0719	.0755	4	8	11	15	19	23	26	30	34
12	.0792	.0828	.0864	.0899	.0934	.0969	.1004	.1038	.1072	.1106	3	7	10	14	17	21	24	28	31
13	.1139	.1173	.1206	.1239	.1271	.1303	.1335	.1367	.1399	.1430	3	6	10	13	16	19	23	26	29
14	.1461	.1492	.1523	.1553	.1584	.1614	.1644	.1673	.1703	.1732	3	6	9	12	15	18	21	24	27
15	.1761	.1790	.1818	.1847	.1875	.1903	.1931	.1959	.1987	.2014	3	6	8	11	14	17	20	22	25
16	.2041	.2068	.2095	.2122	.2148	.2175	.2201	.2227	.2253	.2279	3	5	8	11	13	16	18	21	24
17	.2304	.2330	.2355	.2380	.2405	.2430	.2455	.2480	.2504	.2529	2	5	7	10	12	15	17	20	22
18	.2553	.2577	.2601	.2625	.2648	.2672	.2695	.2718	.2742	.2765	2	5	7	9	12	14	16	19	21
19	.2788	.2810	.2833	.2856	.2878	.2900	.2923	.2945	.2967	.2989	2	4	7	9	11	13	16	18	20
20	.3010	.3032	.3054	.3075	.3096	.3118	.3139	.3160	.3181	.3201	2	4	6	8	11	13	15	17	19
21	.3222	.3243	.3263	.3284	.3304	.3324	.3345	.3365	.3385	.3404	2	4	6	8	10	12	14	16	18
22	.3424	.3444	.3464	.3483	.3502	.3522	.3541	.3560	.3579	.3598	2	4	6	8	10	12	14	15	17
23	.3617	.3636	.3655	.3674	.3692	.3711	.3729	.3747	.3766	.3784	2	4	6	7	9	11	13	15	17
24	.3802	.3820	.3838	.3856	.3874	.3892	.3909	.3927	.3945	.3962	2	4	5	7	9	11	12	14	16
25	.3979	.3997	.4014	.4031	.4048	.4065	.4082	.4099	.4116	.4133	2	3	5	7	9	10	12	14	15
26	.4150	.4166	.4183	.4200	.4216	.4232	.4249	.4265	.4281	.4298	2	3	5	7	8	10	11	13	15
27	.4314	.4330	.4346	.4362	.4378	.4393	.4409	.4425	.4440	.4456	2	3	5	6	8	9	11	13	14
28	.4472	.4487	.4502	.4518	.4533	.4548	.4564	.4579	.4594	.4609	2	3	5	6	8	9	11	12	14
29	.4624	.4639	.4654	.4669	.4683	.4698	.4713	.4728	.4742	.4757	1	3	4	6	7	9	10	12	13
30	.4771	.4786	.4800	.4814	.4829	.4843	.4857	.4871	.4886	.4900	1	3	4	6	7	9	10	11	13
31	.4914	.4928	.4942	.4955	.4969	.4983	.4997	.5011	.5024	.5038	1	3	4	6	7	8	10	11	12
32	.5051	.5065	.5079	.5092	.5105	.5119	.5132	.5145	.5159	.5172	1	3	4	5	7	8	9	11	12
33	.5185	.5198	.5211	.5224	.5237	.5250	.5263	.5276	.5289	.5302	1	3	4	5	6	8	9	10	12
34	.5315	.5328	.5340	.5353	.5366	.5378	.5391	.5403	.5416	.5428	1	3	4	5	6	8	9	10	11
35	.5441	.5453	.5465	.5478	.5490	.5502	.5514	.5527	.5539	.5551	1	2	4	5	6	7	9	10	11
36	.5563	.5575	.5587	.5599	.5611	.5623	.5635	.5647	.5658	.5670	1	2	4	5	6	7	8	10	11
37	.5682	.5694	.5705	.5717	.5729	.5740	.5752	.5763	.5775	.5786	1	2	3	5	6	7	8	9	10
38	.5798	.5809	.5821	.5832	.5843	.5855	.5866	.5877	.5888	.5899	1	2	3	5	6	7	8	9	10
39	.5911	.5922	.5933	.5944	.5955	.5966	.5977	.5988	.5999	.6010	1	2	3	4	5	7	8	9	10
40	.6021	.6031	.6042	.6053	.6064	.6075	.6085	.6096	.6107	.6117	1	2	3	4	5	6	8	9	10
41	.6128	.6138	.6149	.6160	.6170	.6180	.6191	.6201	.6212	.6222	1	2	3	4	5	6	7	8	9
42	.6232	.6243	.6253	.6263	.6274	.6284	.6294	.6304	.6314	.6325	1	2	3	4	5	6	7	8	9
43	.6335	.6345	.6355	.6365	.6375	.6385	.6395	.6405	.6415	.6425	1	2	3	4	5	6	7	8	9
44	.6435	.6444	.6454	.6464	.6474	.6484	.6493	.6503	.6513	.6522	1	2	3	4	5	6	7	8	9
45	.6532	.6542	.6551	.6561	.6571	.6580	.6590	.6599	.6609	.6618	1	2	3	4	5	6	7	8	9
46	.6628	.6637	.6646	.6656	.6665	.6675	.6684	.6693	.6702	.6712	1	2	3	4	5	6	7	7	8
47	.6721	.6730	.6739	.6749	.6758	.6767	.6776	.6785	.6794	.6803	1	2	3	4	5	5	6	7	8
48	.6812	.6821	.6830	.6839	.6848	.6857	.6866	.6875	.6884	.6893	1	2	3	4	4	5	6	7	8
49	.6902	.6911	.6920	.6928	.6937	.6946	.6955	.6964	.6972	.6981	1	2	3	4	4	5	6	7	8
50	.6990	.6998	.7007	.7016	.7024	.7033	.7042	.7050	.7059	.7067	1	2	3	3	4	5	6	7	8
51	.7076	.7084	.7093	.7101	.7110	.7118	.7126	.7135	.7143	.7152	1	2	3	3	4	5	6	7	8
52	.7160	.7168	.7177	.7185	.7193	.7202	.7210	.7218	.7226	.7235	1	2	2	3	4	5	6	6	7
53	.7243	.7251	.7259	.7267	.7275	.7284	.7292	.7300	.7308	.7316	1	2	2	3	4	5	6	6	7
54	.7324	.7332	.7340	.7348	.7356	.7364	.7372	.7380	.7388	.7396	1	2	2	3	4	5	6	6	7

TABLE V. — (Continued).

Nat Nos.	0	1	2	3	4	5	6	7	8	9	Proportional parts.								
											1	2	3	4	5	6	7	8	9
55	.7404	.7412	.7419	.7427	.7435	.7443	.7451	.7459	.7466	.7474	1	2	2	3	4	5	5	6	7
56	.7482	.7490	.7497	.7505	.7513	.7520	.7528	.7536	.7543	.7551	1	2	2	3	4	5	5	6	7
57	.7559	.7566	.7574	.7582	.7589	.7597	.7604	.7612	.7619	.7627	1	2	2	3	4	5	5	6	7
58	.7634	.7642	.7649	.7657	.7664	.7672	.7679	.7686	.7694	.7701	1	1	2	3	4	4	5	6	7
59	.7709	.7716	.7723	.7731	.7738	.7745	.7752	.7760	.7767	.7774	1	1	2	3	4	4	5	6	7
60	.7782	.7789	.7796	.7803	.7810	.7818	.7825	.7832	.7839	.7846	1	1	2	3	4	4	5	6	6
61	.7853	.7860	.7868	.7875	.7882	.7889	.7896	.7903	.7910	.7917	1	1	2	3	4	4	5	6	6
62	.7924	.7931	.7938	.7945	.7952	.7959	.7966	.7973	.7980	.7987	1	1	2	3	4	4	5	6	6
63	.7993	.8000	.8007	.8014	.8021	.8028	.8035	.8041	.8048	.8055	1	1	2	3	3	4	5	5	6
64	.8062	.8069	.8075	.8082	.8089	.8096	.8102	.8109	.8116	.8122	1	1	2	3	3	4	5	5	6
65	.8129	.8136	.8142	.8149	.8156	.8162	.8169	.8176	.8182	.8189	1	1	2	3	3	4	5	5	6
66	.8195	.8202	.8209	.8215	.8222	.8228	.8235	.8241	.8248	.8254	1	1	2	3	3	4	5	5	6
67	.8261	.8267	.8274	.8280	.8287	.8293	.8299	.8306	.8312	.8319	1	1	2	3	3	4	5	5	6
68	.8325	.8331	.8338	.8344	.8351	.8357	.8363	.8370	.8376	.8382	1	1	2	3	3	4	4	5	6
69	.8388	.8395	.8401	.8407	.8414	.8420	.8426	.8432	.8439	.8445	1	1	2	3	4	4	5	6	
70	.8451	.8457	.8463	.8470	.8476	.8482	.8488	.8494	.8500	.8506	1	1	2	2	3	4	4	5	6
71	.8513	.8519	.8525	.8531	.8537	.8543	.8549	.8555	.8561	.8567	1	1	2	2	3	4	4	5	5
72	.8573	.8579	.8585	.8591	.8597	.8603	.8609	.8615	.8621	.8627	1	1	2	2	3	4	4	5	5
73	.8633	.8639	.8645	.8651	.8657	.8663	.8669	.8675	.8681	.8686	1	1	2	2	3	4	4	5	5
74	.8692	.8698	.8704	.8710	.8716	.8722	.8727	.8733	.8739	.8745	1	1	2	2	3	4	4	5	5
75	.8751	.8756	.8762	.8768	.8774	.8779	.8785	.8791	.8797	.8802	1	1	2	2	3	3	4	5	5
76	.8808	.8814	.8820	.8825	.8831	.8837	.8842	.8848	.8854	.8859	1	1	2	2	3	3	4	5	5
77	.8865	.8871	.8876	.8882	.8887	.8893	.8899	.8904	.8910	.8915	1	1	2	2	3	3	4	4	5
78	.8921	.8927	.8932	.8938	.8943	.8949	.8954	.8960	.8965	.8971	1	1	2	2	3	3	4	4	5
79	.8976	.8982	.8987	.8993	.8998	.9004	.9009	.9015	.9020	.9025	1	1	2	2	3	3	4	4	5
80	.9031	.9036	.9042	.9047	.9053	.9058	.9063	.9069	.9074	.9079	1	1	2	2	3	3	4	4	5
81	.9085	.9090	.9096	.9101	.9106	.9112	.9117	.9122	.9128	.9133	1	1	2	2	3	3	4	4	5
82	.9138	.9143	.9149	.9154	.9159	.9165	.9170	.9175	.9180	.9186	1	1	2	2	3	3	4	4	5
83	.9191	.9196	.9201	.9206	.9212	.9217	.9222	.9227	.9232	.9238	1	1	2	2	3	3	4	4	5
84	.9243	.9248	.9253	.9258	.9263	.9269	.9274	.9279	.9284	.9289	1	1	2	2	3	3	4	4	5
85	.9294	.9299	.9304	.9309	.9315	.9320	.9325	.9330	.9335	.9340	1	1	2	2	3	3	4	4	5
86	.9345	.9350	.9355	.9360	.9365	.9370	.9375	.9380	.9385	.9390	1	1	2	2	3	3	4	4	5
87	.9395	.9400	.9405	.9410	.9415	.9420	.9425	.9430	.9435	.9440	0	1	1	2	2	3	3	4	4
88	.9445	.9450	.9455	.9460	.9465	.9469	.9474	.9479	.9484	.9489	0	1	1	2	2	3	3	4	4
89	.9494	.9499	.9504	.9509	.9513	.9518	.9523	.9528	.9533	.9538	0	1	1	2	2	3	3	4	4
90	.9542	.9547	.9552	.9557	.9562	.9566	.9571	.9576	.9581	.9586	0	1	1	2	2	3	3	4	4
91	.9590	.9595	.9600	.9605	.9609	.9614	.9619	.9624	.9628	.9633	0	1	1	2	2	3	3	4	4
92	.9638	.9643	.9647	.9652	.9657	.9661	.9666	.9671	.9675	.9680	0	1	1	2	2	3	3	4	4
93	.9685	.9689	.9694	.9699	.9703	.9708	.9713	.9717	.9722	.9727	0	1	1	2	2	3	3	4	4
94	.9731	.9736	.9741	.9745	.9750	.9754	.9759	.9763	.9768	.9773	0	1	1	2	2	3	3	4	4
95	.9777	.9782	.9786	.9791	.9795	.9800	.9805	.9809	.9814	.9818	0	1	1	2	2	3	3	4	4
96	.9823	.9827	.9832	.9836	.9841	.9845	.9850	.9854	.9859	.9863	0	1	1	2	2	3	3	4	4
97	.9868	.9872	.9877	.9881	.9886	.9890	.9894	.9899	.9903	.9908	0	1	1	2	2	3	3	4	4
98	.9912	.9917	.9921	.9926	.9930	.9934	.9939	.9943	.9948	.9952	0	1	1	2	2	3	3	4	4
99	.9956	.9961	.9965	.9969	.9974	.9978	.9983	.9987	.9991	.9996	0	1	1	2	2	3	3	4	4

TABLE VI. — LOGARITHMIC VERSED SINES.

Deg.	0'	10'	20'	30'	40'	50'	1'	2'	3'	4'	5'	6'	7'	8'	9'
10°	8.1816	8.1959	8.2100	8.2239	8.2375	8.2510	14	27	41	55	69	82	96	110	123
11	.2642	.2772	.2900	.3027	.3151	.3274	12	25	38	50	63	75	88	100	113
12	.3395	.3514	.3632	.3748	.3863	.3976	12	23	35	46	58	69	81	92	104
13	.4087	.4198	.4306	.4414	.4520	.4625	11	21	32	43	53	64	75	85	96
14	.4728	.4830	.4932	.5031	.5130	.5228	10	20	30	40	50	60	70	80	89
15	.5324	.5420	.5514	.5607	.5700	.5791	9	19	28	37	46	56	65	74	84
16	.5881	.5971	.6059	.6147	.6234	.6319	9	17	26	35	44	52	61	70	79
17	.6404	.6488	.6572	.6654	.6736	.6817	8	16	25	33	41	49	57	66	74
18	.6897	.6976	.7055	.7133	.7210	.7287	8	15	23	31	39	46	54	62	70
19	.7362	.7438	.7512	.7586	.7659	.7732	7	15	22	30	37	44	52	59	66
20	.7804	.7875	.7946	.8016	.8086	.8155	7	14	21	28	35	42	49	56	63
21	.8223	.8291	.8358	.8425	.8491	.8557	7	13	20	27	33	40	47	53	60
22	.8622	.8687	.8751	.8815	.8878	.8941	6	13	19	25	32	38	44	51	57
23	.9003	.9065	.9127	.9188	.9248	.9308	6	12	18	24	30	36	43	49	55
24	.9368	.9427	.9486	.9544	.9602	.9660	6	12	17	23	29	35	41	47	52
25	8.9717	8.9774	8.9830	8.9886	8.9942	8.9997	6	11	17	22	28	34	39	45	50
26	9.0052	9.0107	9.0161	9.0215	9.0268	9.0321	5	11	16	21	27	32	38	43	48
27	.0374	.0426	.0479	.0530	.0582	.0633	5	10	16	21	26	31	36	41	47
28	.0684	.0734	.0785	.0834	.0884	.0933	5	10	15	20	25	30	35	40	45
29	.0982	.1031	.1079	.1128	.1175	.1223	5	10	14	19	24	29	34	38	43
30	.1270	.1317	.1364	.1410	.1457	.1503	5	9	14	19	23	28	32	37	42
31	.1548	.1594	.1630	.1684	.1728	.1773	4	9	13	18	22	27	31	36	40
32	.1817	.1861	.1905	.1948	.1991	.2034	4	9	13	17	22	26	30	35	39
33	.2077	.2120	.2162	.2204	.2246	.2288	4	8	13	17	21	25	29	34	38
34	.2329	.2370	.2411	.2452	.2493	.2533	4	8	12	16	20	24	28	33	37
35	.2573	.2613	.2653	.2692	.2732	.2771	4	8	12	16	20	24	28	32	36
36	.2810	.2849	.2887	.2926	.2964	.3002	4	8	11	15	19	23	27	31	34
37	.3040	.3077	.3115	.3152	.3189	.3226	4	7	11	15	19	22	26	30	33
38	.3263	.3300	.3336	.3372	.3409	.3444	4	7	11	14	18	22	25	29	33
39	.3480	.3516	.3551	.3586	.3622	.3657	4	7	11	14	18	21	25	28	32
40	.3691	.3726	.3760	.3795	.3829	.3863	3	7	10	14	17	21	24	27	31
41	.3897	.3931	.3964	.3998	.4031	.4064	3	7	10	13	17	20	23	27	30
42	.4097	.4130	.4162	.4195	.4227	.4260	3	7	10	13	16	20	23	26	29
43	.4292	.4324	.4356	.4387	.4419	.4450	3	6	10	13	16	19	22	25	29
44	.4482	.4513	.4544	.4575	.4606	.4637	3	6	9	12	15	19	22	25	28
45	.4667	.4698	.4728	.4758	.4788	.4818	3	6	9	12	15	18	21	24	27
46	.4848	.4878	.4908	.4937	.4966	.4995	3	6	9	12	15	18	21	24	26
47	.5024	.5053	.5082	.5111	.5140	.5168	3	6	9	11	14	17	20	23	26
48	.5197	.5225	.5253	.5281	.5309	.5337	3	6	8	11	14	17	20	22	25
49	.5365	.5393	.5420	.5448	.5475	.5502	3	5	8	11	14	16	19	22	25
50	.5529	.5556	.5583	.5610	.5637	.5663	3	5	8	11	13	16	19	21	24
51	.5690	.5716	.5743	.5769	.5795	.5821	3	5	8	10	13	16	18	21	24
52	.5847	.5873	.5899	.5924	.5950	.5975	3	5	8	10	13	15	18	21	23
53	.6001	.6026	.6051	.6076	.6101	.6126	3	5	8	10	13	15	18	20	23
54	.6151	.6176	.6201	.6225	.6250	.6274	2	5	7	10	12	15	17	20	22
55	.6298	.6323	.6347	.6371	.6395	.6419	2	5	7	10	12	14	17	19	22
56	.6442	.6466	.6490	.6513	.6537	.6560	2	5	7	9	12	14	16	19	21
57	.6584	.6607	.6630	.6653	.6676	.6699	2	5	7	9	11	14	16	18	21
58	.6722	.6744	.6767	.6790	.6812	.6835	2	5	7	9	11	14	16	18	20
59	.6857	.6879	.6902	.6924	.6946	.6968	2	4	7	9	11	13	15	18	20

TABLE VI. — (Continued).

Deg.	0'	10'	20'	30'	40'	50'	1'	2'	3'	4'	5'	6'	7'	8'	9'
				Minutes.						Proportional parts.					
60°	9.6990	9.7012	9.7033	9.7055	9.7077	9.7008	2	4	6	9	11	13	15	17	20
61	.7120	.7141	.7162	.7184	.7205	.7226	2	4	6	8	11	13	15	17	19
62	.7247	.7268	.7289	.7310	.7331	.7351	2	4	6	8	10	12	15	17	19
63	.7372	.7393	.7413	.7434	.7454	.7474	2	4	6	8	10	12	14	16	18
64	.7494	.7515	.7535	.7555	.7575	.7595	2	4	6	8	10	12	14	16	18
65	.7615	.7634	.7654	.7674	.7693	.7713	2	4	6	8	10	12	14	16	18
66	.7732	.7752	.7771	.7791	.7810	.7829	2	4	6	8	10	12	14	15	17
67	.7848	.7867	.7886	.7905	.7924	.7943	2	4	6	8	9	11	13	15	17
68	.7962	.7980	.7999	.8017	.8036	.8054	2	4	6	7	9	11	13	15	17
69	.8073	.8091	.8110	.8128	.8146	.8164	2	4	5	7	9	11	13	15	16
70	.8182	.8200	.8218	.8236	.8254	.8272	2	4	5	7	9	11	13	14	16
71	.8289	.8307	.8325	.8342	.8360	.8377	2	4	5	7	9	11	12	14	16
72	.8395	.8412	.8429	.8447	.8464	.8481	2	3	5	7	9	10	12	14	15
73	.8498	.8515	.8532	.8549	.8566	.8583	2	3	5	7	8	10	12	14	15
74	.8600	.8616	.8633	.8650	.8666	.8683	2	3	5	7	8	10	12	13	15
75	.8699	.8716	.8732	.8748	.8765	.8781	2	3	5	7	8	10	11	13	15
76	.8797	.8813	.8829	.8845	.8861	.8877	2	3	5	6	8	10	11	13	14
77	.8893	.8909	.8925	.8941	.8956	.8972	2	3	5	6	8	9	11	13	14
78	.8988	.9003	.9019	.9034	.9050	.9065	2	3	5	6	8	9	11	12	14
79	.9081	.9096	.9111	.9126	.9141	.9157	2	3	5	6	8	9	11	12	14
80	.9172	.9187	.9202	.9217	.9232	.9246	1	3	4	6	7	9	10	12	13
81	.9261	.9276	.9291	.9305	.9320	.9335	1	3	4	6	7	9	10	12	13
82	.9349	.9364	.9378	.9393	.9407	.9421	1	3	4	6	7	9	10	12	13
83	.9436	.9450	.9464	.9478	.9492	.9506	1	3	4	6	7	9	10	11	13
84	.9521	.9535	.9548	.9562	.9576	.9590	1	3	4	6	7	8	10	11	13
85	.9604	.9618	.9631	.9645	.9659	.9672	1	3	4	5	7	8	10	11	12
86	.9686	.9699	.9713	.9726	.9740	.9753	1	3	4	5	7	8	9	11	12
87	.9767	.9780	.9793	.9806	.9819	.9833	1	3	4	5	7	8	9	11	12
88	.9846	.9859	.9872	.9885	.9898	.9911	1	3	4	5	6	8	9	10	12
89	9.9924	9.9936	9.9949	9.9962	9.9975	9.9987	1	3	4	5	6	8	9	10	11
90	0.0000	0.0013	0.0025	0.0038	0.0050	0.0063	1	3	4	5	6	8	9	10	11
91	.0075	.0088	.0100	.0112	.0125	.0137	1	2	4	5	6	7	9	10	11
92	.0149	.0161	.0173	.0185	.0197	.0210	1	2	4	5	6	7	9	10	11
93	.0222	.0234	.0245	.0257	.0269	.0281	1	2	4	5	6	7	8	9	11
94	.0293	.0305	.0316	.0328	.0340	.0351	1	2	3	5	6	7	8	9	10
95	.0363	.0374	.0386	.0397	.0409	.0420	1	2	3	5	6	7	8	9	10
96	.0432	.0443	.0454	.0466	.0477	.0488	1	2	3	4	6	7	8	9	10
97	.0499	.0511	.0522	.0533	.0544	.0555	1	2	3	4	6	7	8	9	10
98	.0566	.0577	.0588	.0599	.0610	.0620	1	2	3	4	5	6	8	9	10
99	.0631	.0642	.0653	.0663	.0674	.0685	1	2	3	4	5	6	8	9	10
100	.0695	.0706	.0717	.0727	.0738	.0748	1	2	3	4	5	6	7	8	10
101	.0758	.0769	.0779	.0790	.0800	.0810	1	2	3	4	5	6	7	8	9
102	.0820	.0831	.0841	.0851	.0861	.0871	1	2	3	4	5	6	7	8	9
103	.0881	.0891	.0901	.0911	.0921	.0931	1	2	3	4	5	6	7	8	9
104	.0941	.0951	.0961	.0970	.0980	.0990	1	2	3	4	5	6	7	8	9
105	.1000	.1009	.1019	.1029	.1038	.1048	1	2	3	4	5	6	7	8	9
106	.1057	.1067	.1076	.1086	.1095	.1105	1	2	3	4	5	6	7	8	9
107	.1114	.1123	.1133	.1142	.1151	.1160	1	2	3	4	5	6	6	7	8
108	.1169	.1179	.1188	.1197	.1206	.1215	1	2	3	4	5	5	6	7	8
109	.1224	.1233	.1242	.1251	.1260	.1269	1	2	3	4	5	5	6	7	8

TABLE VII. — HOUR ANGLES OF δ CASSIOPEIÆ.

Intervals: — 1910, 6^m58^s; 1920, 10^m57^s; 1930, 15^m13^s.

	Latitudes.									
Alt.	16°	18°	20°	22°	24°	26°	28°	30°	32°	34°
6°	106°.0	109°.9	113°.8	118°.0	122°.5	127°.3	132°.7	138°.7	145°.9	155°.2
8	101.8	105.5	109.3	113.3	117.5	122.0	126.9	132.2	138.3	145.5
10	97.7	101.2	104.9	108.8	112.8	117.0	121.5	126.4	131.8	137.9
12	93.6	97.1	100.7	104.4	108.2	112.2	116.4	121.0	125.9	131.3
14	89.6	93.0	96.5	100.1	103.8	107.6	111.6	115.9	120.4	125.3
16	85.6	89.0	92.4	95.9	99.5	103.2	107.0	111.0	115.3	119.8
18	81.6	85.0	88.4	91.8	95.3	98.9	102.6	106.4	110.4	114.7
20	77.6	81.0	84.4	87.7	91.2	94.7	98.2	101.9	105.8	109.8
22	73.6	77.0	80.4	83.7	87.1	90.5	94.0	97.6	101.3	105.1
24	69.6	73.0	76.4	79.7	83.1	86.5	89.9	93.3	96.9	100.6
26	65.6	69.0	72.4	75.8	79.1	82.4	85.8	89.2	92.6	96.2
28	61.4	65.0	68.4	71.8	75.1	78.4	81.7	85.1	88.4	91.9
30	57.2	60.9	64.4	67.8	71.2	74.5	77.7	81.0	84.3	87.7
32	52.9	56.7	60.3	63.8	67.2	70.5	73.8	77.0	80.3	83.6
34	48.4	52.4	56.1	59.7	63.2	66.5	69.8	73.0	76.3	79.5
36	43.7	47.9	51.8	55.5	59.1	62.5	65.8	69.1	72.3	75.5
38	38.7	43.2	47.4	51.3	54.9	58.5	61.8	65.1	68.3	71.5
40	33.2	38.3	42.8	46.9	50.7	54.3	57.8	61.1	64.3	67.5
42	27.0	32.9	37.8	42.2	46.3	50.1	53.6	57.0	60.3	63.5
44	......	26.7	32.4	37.3	41.7	45.7	49.4	53.0	56.3	59.6
46	......	......	26.3	32.0	36.9	41.2	45.1	48.8	52.2	55.5
48	......	......	......	26.0	31.6	36.3	40.6	44.5	48.1	51.5
50	......	......	......	......	25.6	31.2	35.8	40.0	43.8	47.3
52	......	......	......	......	......	25.2	30.6	35.2	39.3	43.1
54	......	......	......	......	......	......	24.8	30.1	34.7	38.7

	Latitudes.									
Alt.	36°	38°	40°	42°	44°	46°	48°	50°	52°	54°
8°	154°.9	°	°	°	°	°	°	°	°	°
10	145.2	154.6								
12	137.5	144.8	154.4							
14	130.8	137.0	144.4	154.0						
16	124.8	130.3	136.5	144.0	153.7					
18	119.2	124.2	129.7	136.0	143.6	153.4				
20	114.0	118.6	123.6	129.2	135.5	143.1	153.0			
22	109.1	113.4	118.0	123.0	128.6	135.0	142.6	152.7		
24	104.4	108.4	112.7	117.3	122.3	127.9	134.4	142.1	152.2	
26	99.8	103.7	107.7	112.0	116.6	121.6	127.2	133.7	141.5	151.8
28	95.4	99.1	102.9	106.9	111.2	115.8	120.9	126.5	133.0	140.9
30	91.1	94.6	98.3	102.1	106.1	110.4	115.0	120.0	125.7	132.3
32	86.9	90.3	93.8	97.4	101.2	105.2	109.5	114.1	119.2	124.9
34	82.8	86.1	89.4	92.9	96.5	100.3	104.3	108.5	113.1	118.2
36	78.7	81.9	85.2	88.5	92.0	95.5	99.3	103.3	107.5	112.1
38	74.7	77.8	81.0	84.3	87.6	91.0	94.5	98.2	102.2	106.4
40	70.7	73.8	76.9	80.1	83.3	86.5	89.9	93.4	97.1	101.0
42	66.7	69.8	72.9	75.9	79.0	82.2	85.4	88.7	92.2	95.8
44	62.7	65.8	68.8	71.9	74.9	77.9	81.0	84.3	87.5	90.9
46	58.7	61.8	64.8	67.8	70.8	73.8	76.8	79.8	82.9	86.1
48	54.7	57.8	60.9	63.8	66.8	69.7	72.6	75.5	78.4	81.4
50	50.7	53.8	56.9	59.9	62.8	65.6	68.4	71.2	74.1	76.9
52	46.5	49.8	52.9	55.9	58.8	61.6	64.3	67.1	69.8	72.5
54	42.3	45.7	48.9	51.9	54.8	57.6	60.3	63.0	65.6	68.2
56	37.9	41.5	44.8	47.9	50.8	53.6	56.3	58.9	61.5	63.9
58	33.3	37.2	40.6	43.8	46.8	49.6	52.3	54.9	57.4	59.8
60	28.4	32.6	36.3	39.7	42.8	45.6	48.3	50.9	53.3	55.7

NOTE. — If the star is east of the meridian subtract this hour angle from 360°.

TABLE VIII. — COÖRDINATES OF POLARIS.

H. A.	1910 sin t	1910 cos t	1920 sin t	1920 cos t	1930 sin t	1930 cos t	H. A.	H. A.	1910 sin t	1910 cost	1920 sin t	1920 cost	1930 sin t	1930 cost	H. A.
30	35.2	61.0	33.7	58.3	32.1	55.6	150	60	61.0	35.2	58.3	33.7	55.6	32.1	120
32	37.3	59.7	35.7	57.1	34.1	54.5	148	62	62.2	33.1	59.5	31.6	56.7	30.2	118
34	39.4	58.4	37.7	55.8	35.9	53.3	146	64	63.3	30.9	60.5	29.5	57.8	28.2	116
36	41.4	57.0	39.6	54.5	37.8	52.0	144	66	64.4	28.6	61.5	27.4	58.7	26.1	114
38	43.4	55.5	41.5	53.1	39.6	50.6	142	68	65.3	26.4	62.4	25.2	59.6	24.1	112
40	45.3	54.0	43.3	51.6	41.3	49.2	140	70	66.2	24.1	63.3	23.0	60.4	22.0	110
42	47.1	52.4	45.1	50.0	43.0	47.7	138	72	67.0	21.8	64.1	20.8	61.1	19.9	108
44	48.9	50.7	46.8	48.4	44.6	46.2	136	74	67.7	19.4	64.7	18.6	61.8	17.7	106
46	50.7	48.9	48.4	46.8	46.2	44.6	134	76	68.3	17.0	65.3	16.3	62.3	15.5	104
48	52.4	47.1	50.0	45.1	47.7	43.0	132	78	68.9	14.6	65.9	14.0	62.9	13.4	102
50	54.0	45.3	51.6	43.3	49.2	41.3	130	80	69.4	12.2	66.3	11.7	63.3	11.2	100
52	55.5	43.4	53.1	41.5	50.6	39.6	128	82	69.8	9.8	66.7	9.4	63.6	8.9	98
54	57.0	41.4	54.5	39.6	52.0	37.8	126	84	70.1	7.4	67.0	7.0	63.9	6.7	96
56	58.4	39.4	55.8	37.7	53.3	35.9	124	86	70.3	4.9	67.2	4.7	64.1	4.5	94
58	59.7	37.3	57.1	35.7	54.5	34.1	122	88	70.4	2.5	67.3	2.4	64.2	2.2	92
60	61.0	35.2	58.3	33.7	55.6	32.1	120	90	70.4	0.0	67.4	0.0	64.3	0.0	90

TABLE IX. — CORRECTION FOR AZIMUTH.

Alt.	p sin t.					Proportional parts.								
	30'	40'	50'	60	70'	1'	2'	3'	4'	5'	6'	7'	8'	9'
15°	1.1	1.4	1.8	2.1	2.5	.0	.1	.1	.1	.2	.2	.2	.3	.3
18	1.5	2.1	2.6	3.1	3.6	.1	.1	.2	.2	.3	.3	.4	.4	.5
21	2.1	2.8	3.6	4.3	5.0	.1	.1	.2	.3	.4	.4	.5	.6	.6
24	2.8	3.8	4.7	5.7	6.6	.1	.2	.3	.4	.5	.6	.7	.8	.9
27	3.7	4.9	6.1	7.3	8.6	.1	.2	.4	.5	.6	.7	.9	1.0	1.1
30	4.6	6.2	7.7	9.3	10.8	.2	.3	.5	.6	.8	.9	1.1	1.2	1.4
31	5.0	6.7	8.3	10.0	11.7	.2	.3	.5	.7	.8	1.0	1.2	1.3	1.5
32	5.4	7.2	9.0	10.8	12.5	.2	.4	.5	.7	.9	1.1	1.3	1.4	1.6
33	5.8	7.7	9.6	11.5	13.5	.2	.4	.6	.8	1.0	1.2	1.3	1.5	1.7
34	6.2	8.2	10.3	12.4	14.4	.2	.4	.6	.8	1.0	1.2	1.4	1.6	1.9
35	6.6	8.8	11.0	13.2	15.5	.2	.4	.7	.9	1.1	1.3	1.5	1.8	2.0
36	7.1	9.4	11.8	14.2	16.5	.2	.5	.7	.9	1.2	1.4	1.7	1.9	2.1
37	7.6	10.1	12.6	15.1	17.6	.3	.5	.8	1.0	1.3	1.5	1.8	2.0	2.3
38	8.1	10.8	13.5	16.1	18.8	.3	.5	.8	1.1	1.3	1.6	1.9	2.1	2.4
39	8.6	11.5	14.3	17.2	20.1	.3	.6	.9	1.1	1.4	1.7	2.0	2.3	2.6
40	9.2	12.2	15.3	18.3	21.4	.3	.6	.9	1.2	1.5	1.8	2.1	2.4	2.7
41	9.8	13.0	16.3	19.5	22.8	.3	.7	1.0	1.3	1.6	2.0	2.3	2.6	2.9
42	10.4	13.8	17.3	20.7	24.2	.3	.7	1.0	1.4	1.7	2.1	2.4	2.8	3.1
43	11.0	14.7	18.4	22.0	25.7	.4	.7	1.1	1.5	1.8	2.2	2.6	2.9	3.3
44	11.7	15.6	19.5	23.4	27.3	.4	.8	1.2	1.6	2.0	2.3	2.7	3.1	3.5
45	12.4	16.6	20.7	24.9	29.0	.4	.8	1.2	1.7	2.1	2.5	2.9	3.3	3.7
46	13.2	17.6	22.0	26.4	30.8	.4	.9	1.3	1.8	2.2	2.6	3.1	3.5	3.9
47	14.0	18.7	23.3	28.0	32.6	.5	.9	1.4	1.9	2.3	2.8	3.3	3.7	4.2
48	14.8	19.8	24.7	29.7	34.6	.5	1.0	1.5	2.0	2.5	3.0	3.5	4.0	4.4
49	15.7	21.0	26.2	31.5	36.7	.5	1.0	1.6	2.1	2.6	3.1	3.7	4.1	4.7
50	16.7	22.2	27.8	33.3	38.9	.6	1.1	1.7	2.2	2.8	3.3	3.9	4.4	5.0
51	17.7	23.6	29.5	35.3	41.2	.6	1.2	1.8	2.4	2.9	3.5	4.1	4.7	5.3
52	18.7	25.0	31.2	37.5	43.7	.6	1.2	1.9	2.5	3.1	3.7	4.4	5.0	5.6
53	19.8	26.5	33.1	39.7	46.3	.7	1.3	2.0	2.6	3.3	4.0	4.6	5.3	6.0
54	21.0	28.1	35.1	42.1	49.1	.7	1.4	2.1	2.8	3.5	4.2	4.9	5.6	6.3
55	22.3	29.7	37.2	44.6	52.0	.7	1.5	2.2	3.0	3.7	4.5	5.2	5.9	6.7

TABLE X. — HOUR ANGLES OF δ DRACONIS.

Intervals: 1910, 6ʰ 14ᵐ 21ˢ; 1920, 6ʰ 18ᵐ 58ˢ; 1930, 6ʰ 23ᵐ 54ˢ.

Latitudes.

Alt.	16°	18°	20°	22°	24°	26°	28°	30°	32°	34°
6°	114°.1	119°.8	126°.1	133°.0	140°.6	151°.0	......	......	......	......
8	108.3	113.7	119.5	125.7	132.6	140.7	150.8	......	......	......
10	102.7	107.9	113.3	119.1	125.4	132.3	140.4	150.5	......	......
12	97.3	102.3	107.5	112.9	118.7	125.0	132.0	140.1	......	......
14	92.0	96.9	101.9	107.1	112.5	118.3	124.6	131.6	139.8	......
16	86.7	91.6	96.5	101.5	106.7	112.1	117.9	124.2	131.3	130.5
18	81.5	86.3	91.1	96.0	101.0	106.2	111.6	117.5	123.8	130.9
20	76.3	81.1	85.9	90.7	95.6	100.6	105.8	111.2	117.1	123.4
22	71.0	75.8	80.6	85.4	90.2	95.1	100.1	105.3	110.8	116.6
24	65.6	70.5	75.4	80.2	84.9	89.7	94.6	99.6	104.8	110.3
26	60.0	65.2	70.1	74.9	79.7	84.4	89.2	94.1	99.1	104.3
28	54.2	59.6	64.7	69.6	74.5	79.2	83.9	88.7	93.6	98.6
30	48.1	53.9	59.2	64.3	69.2	74.0	78.7	83.4	88.2	93.0
32	41.5	47.8	53.5	58.8	63.8	68.7	73.5	78.2	82.9	87.6
34	34.1	41.2	47.4	53.1	58.3	63.4	68.2	72.9	77.6	82.3
36	25.0	33.8	40.9	47.0	52.6	57.9	62.9	67.7	72.4	77.0
38	......	24.8	33.5	40.5	46.6	52.2	57.4	62.4	66.9	71.8
40	......	......	24.6	33.2	40.2	46.2	51.8	56.0	61.8	66.6
42	......	......	......	24.4	32.9	39.8	45.8	51.3	56.4	61.3
44	......	......	......	......	24.1	32.6	39.4	45.4	50.8	55.9
46	......	......	......	......	......	23.9	32.3	39.0	44.9	50.3
48	......	......	......	......	......	......	23.6	31.9	38.6	44.4
50	......	......	......	......	......	......	......	......	......	38.1

Latitudes.

Alt.	36°	38°	40°	42°	44°	46°	48°	50°	52°	54°
18°	130°.2	......	......	......	......	......	......	......	......	......
20	130.5	138.8	......	......	......	......	......	......	......	......
22	123.0	130.1	138.5	......	......	......	......	......	......	......
24	116.2	122.5	129.7	138.1	......	......	......	......	......	......
26	109.8	115.7	122.1	129.3	137.7	......	......	......	......	......
28	103.8	109.3	115.1	121.6	128.8	137.3	......	......	......	......
30	98.0	103.2	108.7	114.6	121.0	128.3	136.8	......	......	......
32	92.4	97.4	102.6	108.1	114.0	120.5	127.8	136.4	......	......
34	87.0	91.8	96.8	102.0	107.5	113.4	119.9	127.2	135.8	......
36	81.7	86.4	91.2	96.2	101.3	106.8	112.7	119.2	126.6	135.3
38	76.5	81.1	85.7	90.5	95.5	100.6	106.1	112.0	118.5	125.9
40	71.2	75.8	80.4	85.1	89.8	94.7	99.9	105.3	111.2	117.8
42	66.0	70.6	75.1	79.6	84.3	89.0	93.9	99.1	104.5	110.4
44	60.7	65.3	69.9	74.4	78.9	83.5	88.2	93.1	98.2	103.6
46	55.3	60.0	64.7	69.2	73.6	78.1	82.7	87.3	92.1	97.2
48	49.7	54.7	59.4	63.9	68.4	72.8	77.2	81.7	86.3	91.1
50	43.9	49.1	54.0	58.7	63.2	67.6	71.9	76.3	80.7	85.3
52	37.7	43.3	48.5	53.3	57.9	62.3	66.7	71.0	75.3	79.6
54	......	37.2	42.8	47.8	52.6	57.1	61.4	65.7	69.9	74.1
56	......	......	36.6	42.1	47.1	51.8	56.2	60.5	64.6	68.7
58	......	......	......	36.0	41.4	46.3	50.9	55.2	59.4	63.4
60	......	......	......	......	35.4	40.7	45.5	49.9	54.1	58.2

NOTE.—If the star is east of the meridian subtract this hour angle from 360°.

TABLE XI. — COÖRDINATES OF POLARIS.

H.A.	1910 p sin t	p cos t	1920 p sin t	p cos t	1930 p sin t	p cos t	H.A.	H.A.	1910 p sin t	p cos t	1920 p sin t	p cos t	1930 p sin t	p cos t	H.A.
0	0.0	70.4	0.0	67.4	0.0	64.3	180	15	18.2	68.0	17.4	65.1	16.6	62.1	165
1	1.2	70.4	1.2	67.4	1.1	64.2	179	16	19.4	67.7	18.6	64.7	17.7	61.8	164
2	2.5	70.4	2.4	67.4	2.2	64.2	178	17	20.6	67.4	19.7	64.4	18.8	61.5	163
3	3.7	70.3	3.5	67.3	3.4	64.2	177	18	21.8	67.0	20.8	64.1	19.9	61.1	162
4	4.9	70.2	4.7	67.2	4.5	64.1	176	19	22.9	66.6	21.9	63.7	20.9	60.8	161
5	6.1	70.1	5.9	67.1	5.6	64.0	175	20	24.1	66.2	23.0	63.3	22.0	60.4	160
6	7.4	70.0	7.0	67.0	6.7	63.9	174	21	25.2	65.8	24.1	62.9	23.0	60.0	159
7	8.6	69.9	8.2	66.9	7.8	63.8	173	22	26.4	65.3	25.2	62.4	24.1	59.6	158
8	9.8	69.7	9.4	66.7	8.9	63.6	172	23	27.5	64.8	26.3	62.0	25.1	59.1	157
9	11.0	69.5	10.5	66.5	10.1	63.5	171	24	28.6	64.4	27.4	61.5	26.1	58.7	156
10	12.2	69.3	11.7	66.3	11.2	63.3	170	25	29.8	63.8	28.5	61.0	27.2	58.2	155
11	13.4	69.1	12.8	66.2	12.3	63.1	169	26	30.9	63.3	29.5	60.5	28.2	57.8	154
12	14.6	68.9	14.0	65.9	13.4	62.9	168	27	32.0	62.8	30.6	60.0	29.2	57.3	153
13	15.8	68.6	15.1	65.7	14.4	62.6	167	28	33.1	62.2	31.6	59.5	30.2	56.7	152
14	17.0	68.3	16.3	65.4	15.5	62.3	166	29	34.2	61.6	32.7	58.9	31.2	56.2	151
15	18.2	68.0	17.4	65.1	16.6	62.1	165	30	35.2	61.0	33.7	58.3	32.1	55.6	150

TABLE XII. — CORRECTIONS FOR AZIMUTH.

Alt.	p sin t. 10′	20′	30′	40′	Proportional parts. 1′	2′	3′	4′	5′	6′	7	8′	9′
15°	′.4	′.7	1′.1	1′.4	′.0	′.1	′.1	′.1	′.2	′.2	′.2	′.3	′.3
18	.5	1.0	1.5	2.1	.1	.1	.2	.2	.3	.3	.4	.4	.5
21	.7	1.4	2.1	2.8	.1	.1	.2	.3	.4	.4	.5	.6	.6
24	.9	1.9	2.8	3.8	.1	.2	.3	.4	.5	.6	.7	.8	.9
27	1.2	2.4	3.7	4.9	.1	.2	.4	.5	.6	.7	.9	1.0	1.1
30	1.5	3.1	4.6	6.2	.2	.3	.5	.6	.8	.9	1.1	1.2	1.4
31	1.7	3.3	5.0	6.7	.2	.3	.5	.7	.8	1.0	1.2	1.3	1.5
32	1.8	3.6	5.4	7.2	.2	.4	.5	.7	.9	1.1	1.3	1.4	1.6
33	1.9	3.8	5.8	7.7	.2	.4	.6	.8	1.0	1.2	1.3	1.5	1.7
34	2.1	4.1	6.2	8.2	.2	.4	.6	.8	1.0	1.2	1.4	1.6	1.9
35	2.2	4.4	6.6	8.8	.2	.4	.7	.9	1.1	1.3	1.5	1.8	2.0
36	2.4	4.7	7.1	9.4	.2	.5	.7	.9	1.2	1.4	1.7	1.9	2.1
37	2.5	5.0	7.6	10.1	.3	.5	.8	1.0	1.3	1.5	1.8	2.0	2.3
38	2.7	5.4	8.1	10.8	.3	.5	.8	1.1	1.3	1.6	1.9	2.1	2.4
39	2.9	5.7	8.6	11.5	.3	.6	.9	1.1	1.4	1.7	2.0	2.3	2.6
40	3.1	6.1	9.2	12.2	.3	.6	.9	1.2	1.5	1.8	2.1	2.4	2.7
41	3.3	6.5	9.8	13.0	.3	.7	1.0	1.3	1.6	2.0	2.3	2.6	2.9
42	3.5	6.9	10.4	13.8	.3	.7	1.0	1.4	1.7	2.1	2.4	2.8	3.1
43	3.7	7.3	11.0	14.7	.4	.7	1.1	1.5	1.8	2.2	2.6	2.9	3.3
44	3.9	7.8	11.7	15.6	.4	.8	1.2	1.6	2.0	2.3	2.7	3.1	3.5
45	4.1	8.3	12.4	16.6	.4	.8	1.2	1.7	2.1	2.5	2.9	3.3	3.7
46	4.4	8.8	13.2	17.6	.4	.9	1.3	1.8	2.2	2.6	3.1	3.5	3.9
47	4.7	9.3	14.0	18.7	.5	.9	1.4	1.9	2.3	2.8	3.3	3.7	4.2
48	4.9	9.9	14.8	19.8	.5	1.0	1.5	2.0	2.5	3.0	3.5	4.0	4.4
49	5.2	10.5	15.7	21.0	.5	1.0	1.6	2.1	2.6	3.1	3.7	4.1	4.7
50	5.6	11.1	16.7	22.2	.6	1.1	1.7	2.2	2.8	3.3	3.9	4.4	5.0
51	5.9	11.8	17.7	23.6	.6	1.2	1.8	2.4	2.9	3.5	4.1	4.7	5.3
52	6.2	12.5	18.7	25.0	.6	1.2	1.9	2.5	3.1	3.7	4.4	5.0	5.6
53	6.6	13.2	19.8	26.5	.7	1.3	2.0	2.6	3.3	4.0	4.6	5.3	6.0
54	7.0	14.0	21.0	28.1	.7	1.4	2.1	2.8	3.5	4.2	4.9	5.6	6.3
55	7.4	14.9	22.3	29.7	.7	1.5	2.2	3.0	3.7	4.5	5.2	5.9	6.7

TABLE XIII. — VALUES OF K
FOR COMPUTING THE
ALTITUDE.

$\frac{1}{2} p^2 \sin 1' \sin^2 t \tan h$ for $p = 1° 10'$

TABLE XIII (A).

	Latitudes.		
Hour angle	30°	40°	50°
0°	.0	.0	.0
15	.0	.0	.1
30	.1	.2	.2
45	.2	.3	.4
60	.3	.5	.7
75	.4	.6	.8
90	.4	.6	.9
105	.4	.6	.8
120	.3	.4	.6
135	.2	.3	.4
150	.1	.1	.2
165	.0	.0	.1
180	.0	.0	.0

Equation of time.	
April 15, 0m	May 1, −3m.0
May 15, −3m.8	June 1, −2m.5
June 15, 0m	July 1, +3m.5
July 26 +6m.3	Aug. 1, +6m.1
Aug. 15, +4m.4	Sept. 1, 0m
Sept. 15, −4m.8	Oct. 1, −10m.2
Oct. 15, −14m.1	Nov. 3, −16m.3
Nov. 15, −15m.3	Dec. 1, −10m.9
Dec. 24, 0m	Jan. 1, +3m.2
Jan. 15, +9m.2	Feb. 1, +13m.6
Feb. 12, +14m.4	Mar. 1, +12m.5
Mar. 15, +9m.1	April 1, +4m.0

For an increase of 1′ in p the term increases about 3 %.

TABLE XIII (B).

Sines of Azimuth and Hour Angle.

	0°	1°	2°	3°	4°	5°	6°	7°	8°	9°	10°	
0°	.....	.....	.....	.....	.....	.....	.019	.086	.144	.194	.240	80°
10	.240	.281	.318	.352	.384	.413	.440	.466	.490	.513	.534	70
20	.534	.554	.574	.592	.609	.626	.642	.657	.672	.686	.699	60
30	.699	.712	.724	.736	.748	.759	.769	.779	.789	.799	.808	50
40	.808	.817	.826	.834	.842	.849	.857	.864	.871	.878	.884	40
50	.884	.891	.897	.902	.908	.913	.919	.924	.928	.933	.938	30
60	.938	.942	.946	.950	.954	.957	.961	.964	.967	.970	.973	20
70	.973	.976	.978	.981	.983	.985	.987	.989	.990	.992	.993	10
80	.993	.995	.996	.997	.998	.998	.999	.999	.000	.000	.000	0
	10°	9°	8°	7°	6°	5°	4°	3°	2°	1°	0°	

Cosines of Declination and Altitude.

TABLE XIV. — FOR FINDING THE CORRECTION TO THE SUN'S DECLINATION. (In minutes and tenths.)

Hours from Greenwich Noon	Difference for 1 hour.					Proportional parts.								
	10″	20″	30″	40″	50″	1″	2″	3″	4″	5″	6″	7″	8″	9″
0	.0	.1	.1	.2	.2	.0	.0	.0	.0	.0	.0	.0	.0	.0
	.1	.2	.3	.3	.4	.0	.0	.0	.0	.0	.1	.1	.1	.1
	.1	.3	.4	.5	.6	.0	.0	.0	.1	.1	.1	.1	.1	.1
1	.2	.3	.5	.7	.8	.0	.0	.1	.1	.1	.1	.1	.1	.2
	.2	.4	.6	.8	1.0	.0	.0	.1	.1	.1	.1	.2	.2	.2
	.3	.5	.7	1.0	1.3	.0	.1	.1	.1	.1	.2	.2	.2	.2
	.3	.6	.9	1.2	1.5	.0	.1	.1	.1	.2	.2	.2	.2	.3
2	.3	.7	1.0	1.3	1.7	.0	.1	.1	.1	.2	.2	.2	.3	.3
	.4	.8	1.1	1.5	1.9	.0	.1	.1	.2	.2	.2	.3	.3	.3
	.4	.8	1.3	1.7	2.1	.0	.1	.1	.2	.2	.3	.3	.3	.4
	.5	.9	1.4	1.8	2.3	.1	.1	.1	.2	.2	.3	.3	.4	.4
3	.5	1.0	1.5	2.0	2.5	.1	.1	.2	.2	.3	.3	.4	.4	.5
	.5	1.1	1.6	2.2	2.7	.1	.1	.2	.2	.3	.3	.4	.4	.5
	.6	1.2	1.8	2.3	2.9	.1	.1	.2	.2	.3	.4	.4	.5	.5
	.6	1.3	1.9	2.5	3.1	.1	.1	.2	.3	.3	.4	.4	.5	.6
4	.7	1.3	2.0	2.7	3.3	.1	.1	.2	.3	.3	.4	.5	.5	.6
	.7	1.4	2.1	2.8	3.5	.1	.1	.2	.3	.4	.4	.5	.6	.6
	.8	1.5	2.3	3.0	3.8	.1	.2	.2	.3	.4	.5	.5	.6	.7
	.8	1.6	2.4	3.2	4.0	.1	.2	.2	.3	.4	.5	.6	.6	.7
5	.8	1.7	2.5	3.3	4.2	.1	.2	.3	.3	.4	.5	.6	.7	.8
	.9	1.8	2.6	3.5	4.4	.1	.2	.3	.4	.4	.5	.6	.7	.8
	.9	1.8	2.8	3.7	4.6	.1	.2	.3	.4	.5	.6	.6	.7	.8
	1.0	1.9	2.9	3.8	4.8	.1	.2	.3	.4	.5	.6	.7	.8	.9
6	1.0	2.0	3.0	4.0	5.0	.1	.2	.3	.4	.5	.6	.7	.8	.9
	1.0	2.1	3.1	4.2	5.2	.1	.2	.3	.4	.5	.6	.7	.8	.9
	1.1	2.2	3.3	4.3	5.4	.1	.2	.3	.4	.5	.7	.8	.9	1.0
	1.1	2.3	3.4	4.5	5.6	.1	.2	.3	.5	.6	.7	.8	.9	1.0
7	1.2	2.3	3.5	4.7	5.8	.1	.2	.4	.5	.6	.7	.8	.9	1.1
	1.2	2.4	3.6	4.8	6.0	.1	.2	.4	.5	.6	.7	.9	1.0	1.1
	1.3	2.5	3.8	5.0	6.3	.1	.3	.4	.5	.6	.8	.9	1.0	1.1
	1.3	2.6	3.9	5.2	6.5	.1	.3	.4	.5	.7	.8	.9	1.0	1.2
8	1.3	2.7	4.0	5.3	6.7	.1	.3	.4	.5	.7	.8	.9	1.1	1.2
	1.4	2.8	4.1	5.5	6.9	.1	.3	.4	.6	.7	.8	1.0	1.1	1.2
	1.4	2.8	4.3	5.7	7.1	.1	.3	.4	.6	.7	.9	1.0	1.1	1.3
	1.5	2.9	4.4	5.8	7.3	.2	.3	.4	.6	.7	.9	1.0	1.2	1.3
9	1.5	3.0	4.5	6.0	7.5	.2	.3	.5	.6	.8	.9	1.1	1.2	1.4
	1.5	3.1	4.6	6.2	7.7	.2	.3	.5	.6	.8	.9	1.1	1.2	1.4
	1.6	3.2	4.8	6.3	7.9	.2	.3	.5	.6	.8	1.0	1.1	1.3	1.4
	1.6	3.3	4.9	6.5	8.1	.2	.3	.5	.7	.8	1.0	1.1	1.3	1.5
10	1.7	3.3	5.0	6.7	8.3	.2	.3	.5	.7	.8	1.0	1.2	1.3	1.5
	1.7	3.4	5.1	6.8	8.5	.2	.3	.5	.7	.9	1.0	1.2	1.4	1.5
	1.8	3.5	5.3	7.0	8.8	.2	.4	.5	.7	.9	1.1	1.2	1.4	1.6
	1.8	3.6	5.4	7.2	9.0	.2	.4	.5	.7	.9	1.1	1.3	1.4	1.6
11	1.8	3.7	5.5	7.3	9.2	.2	.4	.6	.7	.9	1.1	1.3	1.5	1.7
	1.9	3.8	5.6	7.5	9.4	.2	.4	.6	.8	.9	1.1	1.3	1.5	1.7
	1.9	3.8	5.8	7.7	9.6	.2	.4	.6	.8	1.0	1.2	1.3	1.5	1.7
	2.0	3.9	5.9	7.8	9.8	.2	.4	.6	.8	1.0	1.2	1.4	1.6	1.8
12	2.0	4.0	6.0	8.0	10.0	.2	.4	.6	.8	1.0	1.2	1.4	1.6	1.8

TABLE XV.—SUN'S DECLINATION AT GREENWICH MEAN NOON.
— 1909 —

Day of month	January Declination	January Diff. for 1 hour	February Declination	February Diff. for 1 hour	March Declination	March Diff. for 1 hour	April Declination	April Diff. for 1 hour	May Declination	May Diff. for 1 hour	June Declination	June Diff. for 1 hour
1	S 23° 02'.3	+12".0	S 17° 11'.8	+42".4	S 7° 43'.0	+56".9	N 4° 24'.2	+57".9	N 14° 58'.0	+45".6	N 22° 00'.5	+20".8
2	22 57.3	13.2	16 54.7	43.2	7 20.2	57.2	4 47.3	57.7	15 16.1	45.0	22 08.6	19.8
3	22 51.8	14.3	16 37.3	43.9	6 57.3	57.4	5 10.4	57.5	15 34.0	44.3	22 16.3	18.8
4	22 45.8	15.4	16 19.6	44.6	6 34.2	57.6	5 33.3	57.3	15 51.6	43.7	22 23.7	17.8
5	22 39.4	16.6	16 01.6	45.3	6 11.2	57.9	5 56.2	57.0	16 08.9	43.0	22 30.6	16.9
6	22 32.6	+17.7	15 43.3	+46.0	5 48.0	+58.1	6 18.9	+56.7	16 26.0	+42.4	22 37.2	+15.9
7	22 25.3	18.8	15 24.8	46.7	5 24.7	58.3	6 41.6	56.5	16 42.8	41.7	22 43.3	14.9
8	22 17.6	19.9	15 06.0	47.3	5 01.4	58.4	7 04.1	56.2	16 59.3	41.0	22 49.1	13.9
9	22 09.4	21.0	14 46.9	47.9	4 38.0	58.6	7 26.5	55.9	17 15.5	40.2	22 54.4	12.9
10	22 00.8	22.0	14 27.6	48.6	4 14.5	58.7	7 48.8	55.6	17 31.5	39.5	22 59.4	11.9
11	21 51.8	+23.1	14 08.1	+49.1	3 51.0	+58.8	8 10.9	+55.2	17 47.1	+38.8	23 04.0	+10.9
12	21 42.3	24.2	13 48.3	49.7	3 27.5	58.9	8 32.9	54.9	18 02.5	38.0	23 08.1	9.9
13	21 32.4	25.2	13 28.3	50.3	3 03.9	59.0	8 54.8	54.5	18 17.6	37.3	23 11.9	8.9
14	21 22.1	26.2	13 08.1	50.8	2 40.2	59.1	9 16.5	54.1	18 32.3	36.5	23 15.2	7.8
15	21 11.4	27.3	12 47.7	51.4	2 16.6	59.2	9 38.1	53.8	18 46.8	35.7	23 18.1	6.8
16	21 00.3	+28.3	12 27.0	+51.9	1 52.9	+59.2	9 59.5	+53.4	19 00.9	+34.9	23 20.6	+5.8
17	20 48.8	29.3	12 06.2	52.4	1 29.2	59.3	10 20.8	53.0	19 14.7	34.1	23 22.7	4.7
18	20 36.0	30.3	11 45.1	52.8	1 05.5	59.3	10 41.9	52.5	19 28.2	33.3	23 24.4	3.7
19	20 24.6	31.2	11 23.9	53.3	0 41.7	59.3	11 02.8	52.1	19 41.4	32.5	23 25.7	2.7
20	20 11.9	32.2	11 02.5	53.7	S 0 18.0	59.3	11 23.5	51.6	19 54.2	31.6	23 26.6	1.6
21	19 58.8	+33.1	10 40.9	+54.2	N 0 05.7	+59.3	11 44.1	+51.1	20 06.7	+30.8	23 27.0	+0.6
22	19 45.4	34.1	10 19.2	54.6	0 29.4	59.2	12 04.4	50.6	20 18.9	29.9	23 27.1	−0.4
23	19 31.6	35.0	9 57.3	54.9	0 53.1	59.2	12 24.6	50.0	20 30.7	29.0	23 26.7	1.5
24	19 17.4	35.9	9 35.2	55.3	1 16.7	59.1	12 44.5	49.6	20 42.1	28.2	23 25.9	2.5
25	19 02.9	36.8	9 13.0	55.7	1 40.3	59.0	13 04.3	49.1	20 53.2	27.3	23 24.7	3.5
26	18 48.0	+37.6	8 50.7	+56.0	2 03.9	+59.0	13 23.8	+48.5	21 04.5	+26.4	23 23.1	−4.6
27	18 32.8	38.5	8 28.3	56.3	2 27.5	58.8	13 43.1	48.0	21 14.3	25.5	23 21.1	5.6
28	18 17.2	39.3	S 8 05.7	+56.6	2 50.9	58.6	14 02.0	47.4	21 24.3	24.5	23 18.6	6.6
29	18 01.4	40.1	....	....	3 14.4	58.5	14 21.0	46.8	21 33.9	23.6	23 15.8	7.6
30	17 45.2	40.9	....	....	3 37.7	58.3	N 14 39.6	+46.2	21 43.1	23.6?	N 23 12.5	−8.7
31	S 17 28.6	+41.7	....	....	N 4 01.0	+58.1	....	....	N 21 52.0	+21.7	....	....

TABLE XV. — (1909 Continued.)

Day of month	July Declination	July Diff. for 1 hour	August Declination	August Diff. for 1 hour	September Declination	September Diff. for 1 hour	October Declination	October Diff. for 1 hour	November Declination	November Diff. for 1 hour	December Declination	December Diff. for 1 hour
1	N 23° 08.9	−9.7	N 18° 07.9	−37.5	N 8° 25.9	−54.3	S 3° 02.4	−58.3	S 14° 19.3	−48.3	S 21° 45.6	−23.7
2	23 04.8	10.7	17 52.7	38.2	8 04.1	54.6	3 25.7	58.2	14 38.5	47.7	21 54.9	22.6
3	23 00.3	11.7	17 37.3	38.8	7 42.1	55.0	3 49.0	58.1	14 57.5	47.1	22 03.7	21.6
4	22 55.5	12.7	17 21.6	39.6	7 20.1	55.3	4 12.2	58.0	15 16.2	46.5	22 12.2	20.5
5	22 50.2	13.7	17 05.6	40.3	6 57.9	55.6	4 35.3	57.8	15 34.7	45.9	22 20.2	19.4
6	22 44.5	−14.7	16 49.4	−41.0	6 35.7	−55.8	4 58.4	−57.7	15 52.9	−45.2	22 27.7	−18.4
7	22 38.5	15.6	16 32.8	41.7	6 13.3	56.1	5 21.5	57.5	16 10.9	44.6	22 34.9	17.3
8	22 32.0	16.6	16 16.0	42.4	5 50.8	56.4	5 44.5	57.4	16 28.6	43.9	22 41.5	16.2
9	22 25.2	17.6	15 58.9	43.0	5 28.2	56.6	6 07.4	57.2	16 46.0	43.2	22 47.8	15.0
10	22 18.0	18.6	15 41.6	43.7	5 05.5	56.8	6 30.2	57.0	17 03.1	42.5	22 53.6	13.9
11	22 10.3	−19.5	15 24.0	−44.3	4 42.7	−57.1	6 53.0	−56.8	17 20.0	−41.7	22 58.9	−12.8
12	22 02.4	20.5	15 06.2	44.9	4 19.0	57.3	7 15.6	56.5	17 36.5	41.0	23 03.8	11.6
13	21 54.0	21.4	14 48.1	45.5	3 56.0	57.4	7 38.2	56.3	17 52.8	40.2	23 08.2	10.5
14	21 45.2	22.3	14 29.8	46.1	3 33.9	57.6	8 00.7	56.0	18 08.7	39.4	23 12.2	9.3
15	21 36.1	23.3	14 11.3	46.6	3 10.8	57.8	8 23.0	55.7	18 24.3	38.6	23 15.7	8.2
16	21 26.6	−24.2	13 52.5	−47.2	2 47.7	−57.9	8 45.2	−55.4	18 39.6	−37.8	23 18.7	−7.0
17	21 16.8	25.1	13 33.5	47.8	2 24.5	58.0	9 07.3	55.1	18 54.5	37.0	23 21.3	5.8
18	21 06.5	26.0	13 14.3	48.3	2 01.3	58.1	9 29.3	54.8	19 09.2	36.1	23 23.4	4.7
19	20 56.0	26.9	12 54.9	48.8	1 38.0	58.2	9 51.1	54.4	19 23.4	35.2	23 25.1	3.5
20	20 45.0	27.8	12 35.3	49.3	1 14.7	58.3	10 12.8	54.0	19 37.3	34.4	23 26.2	2.3
21	20 33.8	−28.6	12 15.4	−49.8	0 51.4	−58.4	10 34.4	−53.6	19 50.9	−33.4	23 26.9	−1.1
22	20 22.1	29.5	11 55.4	50.3	0 28.0	58.4	10 55.7	53.2	20 04.1	32.5	23 27.1	+0.0
23	20 10.2	30.3	11 35.2	50.7	N 0 04.6	58.5	11 16.9	52.8	20 16.9	31.6	23 26.9	1.2
24	19 57.9	31.2	11 14.8	51.2	S 0 18.8	58.5	11 38.0	52.4	20 29.4	30.7	23 26.2	2.4
25	19 45.2	32.0	10 54.3	51.6	0 42.2	58.5	11 58.8	51.9	20 41.5	29.7	23 25.0	3.6
26	19 32.3	−32.8	10 33.3	−52.0	1 05.6	−58.5	12 19.5	−51.4	20 53.1	−28.7	23 23.3	+4.7
27	19 19.0	33.6	10 12.6	52.5	1 29.0	58.5	12 40.0	51.0	21 04.4	27.8	23 21.2	5.9
28	19 05.4	34.4	9 51.6	52.8	1 52.4	58.4	13 00.3	50.5	21 15.3	26.8	23 18.6	7.1
29	18 51.5	35.2	9 30.4	53.2	2 15.7	58.4	13 20.4	50.0	21 25.8	25.7	23 15.5	8.3
30	18 37.2	36.0	9 09.0	53.6	S 2 39.1	58.3	13 40.2	49.4	21 35.9	24.7	23 12.0	9.4
31	N 18 22.7	−36.7	N 8 47.5	54.0	.....	.....	S 13 59.9	−48.9	.....	.....	S 23 08.0	+10.6

XV.—SUN'S DECLINATION AT GREENWICH MEAN NOON.
— 1910 —

Day of month	January Declination	January Diff. for 1 hour	February Declination	February Diff. for 1 hour	March Declination	March Diff. for 1 hour	April Declination	April Diff. for 1 hour	May Declination	May Diff. for 1 hour	June Declination	June Diff. for 1 hour
1	S 23° 03'.5	+11".7	S 17° 16'.0	+42".3	S 7° 48'.6	+56".8	N 4° 18'.5	+58".0	N 14° 53'.5	+45".8	N 21° 58'.3	+21".0
2	22 58.6	12.9	16 59.0	43.0	7 25.9	57.1	4 41.6	57.8	15 11.7	45.1	22 06.7	20.0
3	22 53.2	14.0	16 41.6	43.7	7 03.0	57.3	5 04.7	57.6	15 29.6	44.5	22 14.5	19.0
4	22 47.4	15.2	16 24.0	44.5	6 40.0	57.6	5 27.7	57.3	15 47.3	43.9	22 22.0	18.1
5	22 41.1	16.3	16 06.1	45.2	6 16.9	57.8	5 50.6	57.1	16 04.7	43.2	22 29.0	17.1
6	22 34.3	+17.4	15 47.9	+45.9	5 53.7	+58.0	6 13.3	+56.8	16 21.9	+42.5	22 35.7	+16.1
7	22 27.2	18.5	15 29.4	46.5	5 30.5	58.2	6 36.0	56.6	16 38.8	41.9	22 41.9	15.2
8	22 19.6	19.6	15 10.6	47.2	5 07.1	58.4	6 58.6	56.3	16 55.4	41.2	22 47.8	14.2
9	22 11.5	20.7	14 51.7	47.8	4 43.8	58.6	7 21.1	56.0	17 11.7	40.5	22 53.3	13.1
10	22 03.0	21.8	14 32.4	48.4	4 20.3	58.7	7 43.4	55.7	17 27.7	39.7	22 58.3	12.1
11	21 54.0	+22.9	14 12.9	+49.0	3 56.8	+58.9	8 05.6	+55.3	17 43.5	+39.0	23 03.0	+11.1
12	21 44.7	23.9	13 53.2	49.6	3 33.2	59.0	8 27.7	55.0	17 58.9	38.2	23 07.2	10.1
13	21 34.9	25.0	13 33.2	50.2	3 09.6	59.1	8 49.6	54.6	18 14.1	37.5	23 11.1	9.1
14	21 24.7	26.0	13 13.0	50.7	2 46.0	59.1	9 11.4	54.2	18 28.9	36.7	23 14.5	8.1
15	21 14.1	27.1	12 52.7	51.2	2 22.3	59.2	9 33.0	53.9	18 43.4	35.9	23 17.5	7.0
16	21 03.1	+28.1	12 32.1	+51.7	1 58.6	+59.3	9 54.4	+53.4	18 57.6	+35.1	23 20.1	+6.0
17	20 51.6	29.1	12 11.3	52.2	1 34.9	59.3	10 15.7	53.0	19 11.5	34.3	23 22.3	5.0
18	20 39.8	30.0	11 50.3	52.7	1 11.2	59.3	10 36.9	52.6	19 25.1	33.5	23 24.1	4.0
19	20 27.6	31.0	11 29.1	53.2	0 47.5	59.3	10 57.8	52.1	19 38.3	32.7	23 25.5	2.9
20	20 15.0	32.0	11 07.7	53.6	S 0 23.8	59.3	11 18.6	51.7	19 51.2	31.8	23 26.4	1.9
21	20 02.1	+32.9	10 46.2	+54.0	0 00.0	+59.2	11 39.2	+51.2	20 03.8	+31.0	23 27.0	+0.8
22	19 48.7	33.8	10 24.5	54.4	N 0 23.6	59.2	11 59.5	50.7	20 16.0	30.1	23 27.1	— 0.2
23	19 35.0	34.8	10 02.7	54.8	0 47.3	59.1	12 19.7	50.2	20 27.9	29.2	23 26.8	1.2
24	19 20.9	35.6	9 40.7	55.2	1 10.9	59.1	12 39.7	49.7	20 39.4	28.4	23 26.1	2.3
25	19 06.5	36.5	9 18.5	55.5	1 34.6	59.0	12 59.5	49.2	20 50.6	27.5	23 25.0	3.3
26	18 51.7	+37.4	8 56.8	+55.9	1 58.1	+58.9	13 19.1	+48.6	21 01.4	+26.6	23 23.5	— 4.3
27	18 36.6	38.2	8 33.8	56.2	2 21.7	58.8	13 38.4	48.1	21 11.8	25.7	23 21.6	5.3
28	18 21.1	39.1	S 8 11.5	+56.5	2 45.1	58.6	13 57.5	47.5	21 21.9	24.7	23 19.3	6.4
29	18 05.3	39.9	....	....	3 08.6	58.5	14 16.4	46.9	21 31.6	23.8	23 16.5	7.4
30	17 49.2	40.7	....	....	3 31.9	58.3	N 14 35.1	+46.4	21 40.9	22.9	N 23 13.4	— 8.4
31	S 17 32.8	+41.5	....	....	N 3 55.2	+58.2	....	....	N 21 49.9	+21.9	....	....

TABLE XV.—(1910 Continued).

Day of month.	July Declination.	July Diff. for 1 hour.	August Declination.	August Diff. for 1 hour.	September Declination.	September Diff. for 1 hour.	October Declination.	October Diff. for 1 hour.	November Declination.	November Diff. for 1 hour.	December Declination.	December Diff. for 1 hour.
1	N 23° 09.8	−9'.4	N 18° 11.4	−37'.3	N 8° 31.0	−54'.3	S 2° 56.9	−58'.3	S 14° 14.7	−48'.5	S 21° 43.4	−24'.0
2	23 05.8	10.4	17 56.4	38.0	8 09.2	54.6	3 20.2	58.3	14 34.0	47.9	21 52.8	22.9
3	23 01.5	11.4	17 41.0	38.9	7 47.3	54.9	3 43.5	58.2	14 53.0	47.3	22 01.7	21.9
4	22 56.7	12.4	17 25.4	39.5	7 25.3	55.2	4 06.7	58.0	15 11.8	46.7	22 10.3	20.8
5	22 51.5	13.4	17 09.4	40.2	7 03.2	55.5	4 29.9	57.9	15 30.4	46.0	22 18.4	19.7
6	22 45.9	−14.4	16 53.2	−40.9	6 40.9	−55.8	4 53.0	−57.8	15 48.7	−45.4	22 26.0	−18.6
7	22 40.0	15.4	16 36.7	41.6	6 18.5	56.1	5 16.1	57.6	16 06.7	44.7	22 33.3	17.5
8	22 33.6	16.4	16 20.0	42.2	5 56.0	56.3	5 39.1	57.4	16 24.5	44.0	22 40.0	16.4
9	22 26.9	17.4	16 03.0	42.9	5 33.5	56.6	6 02.0	57.2	16 41.9	43.3	22 46.4	15.3
10	22 19.7	18.3	15 45.7	43.5	5 10.8	56.8	6 24.9	57.0	16 59.1	42.6	22 52.3	14.2
11	22 12.2	−19.3	15 28.2	−44.1	4 48.0	−57.0	6 47.7	−56.8	17 16.0	−41.9	22 57.7	−13.0
12	22 04.3	20.3	15 10.4	44.7	4 25.2	57.2	7 10.3	56.6	17 32.6	41.1	23 02.7	11.9
13	21 56.0	21.2	14 52.4	45.3	4 02.3	57.4	7 32.9	56.3	17 48.9	40.4	23 07.2	10.7
14	21 47.3	22.1	14 34.1	45.9	3 39.3	57.5	7 55.4	56.0	18 04.9	39.6	23 11.3	9.6
15	21 38.3	23.1	14 15.6	46.5	3 16.3	57.7	8 17.7	55.7	18 20.6	38.8	23 14.9	8.4
16	21 28.9	−24.0	13 56.9	−47.1	2 53.2	−57.8	8 40.0	−55.4	18 35.9	−38.0	23 18.1	−7.3
17	21 19.1	24.9	13 38.0	47.6	2 30.0	58.0	9 02.1	55.1	18 51.0	37.1	23 20.7	6.1
18	21 09.0	25.8	13 18.8	48.1	2 06.8	58.1	9 24.1	54.8	19 05.7	36.3	23 23.0	5.0
19	20 58.5	26.7	12 59.5	48.6	1 43.6	58.2	9 45.9	54.4	19 20.0	35.4	23 24.7	3.8
20	20 47.7	27.5	12 39.9	49.2	1 20.3	58.3	10 07.6	54.1	19 34.0	34.5	23 26.0	2.6
21	20 36.5	−28.4	12 20.2	−49.7	0 57.0	−58.3	10 29.2	−53.7	19 47.6	−33.7	23 26.8	−1.4
22	20 24.9	29.3	12 00.2	50.1	0 33.6	58.4	10 50.6	53.3	20 00.8	32.8	23 27.1	0.3
23	20 13.1	30.1	11 40.1	50.6	N 0 10.3	58.4	11 11.8	52.9	20 13.8	31.8	23 27.0	+0.9
24	20 00.8	31.0	11 19.7	51.1	S 0 13.1	58.5	11 32.9	52.5	20 26.4	30.9	23 26.4	2.1
25	19 48.3	31.8	10 59.2	51.5	0 36.5	58.5	11 53.8	52.0	20 38.6	30.0	23 25.3	3.3
26	19 35.4	−32.6	10 38.5	−51.9	S 0 59.9	−58.5	12 14.5	−51.6	20 50.4	−29.0	23 23.8	+4.5
27	19 22.2	33.4	10 17.7	52.4	1 23.3	58.5	12 35.1	51.1	21 01.8	28.0	23 21.8	5.6
28	19 08.7	34.2	09 56.7	52.8	1 46.7	58.5	12 55.6	50.6	21 12.8	27.0	23 19.3	6.8
29	18 54.8	35.0	09 35.5	53.2	2 10.1	58.5	13 15.6	50.1	21 23.4	26.0	23 16.3	8.0
30	18 40.7	35.8	09 14.1	53.5	S 2 33.5	58.4	13 35.5	49.6	S 21 33.6	25.0	23 12.9	9.2
31	N 18 26.2	−36.5	N 8 52.6	−53.9	......	......	S 13 55.2	−49.0	......	......	S 23 09.0	+10.3

TABLE XV.—SUN'S DECLINATION AT GREENWICH MEAN NOON.
— 1911 —

Day of month	January		February		March		April		May		June	
	Declination	Diff. for 1 hour	Declination	Diff. for 1 hour	Declination	Diff. for 1 hour	Declination	Diff. for 1 hour	Declination	Diff. for 1 hour	Declination	Diff. for 1 hour
1	S 23° 04.7	+11″.5	S 17° 20′.1	+42″.1	S 7° 54.1	+56″.8	N 4° 12.9	+58″.1	N 14° 49.2	+45″.9	N 21° 56.6	+21″.2
2	22 59.8	12.6	17 03.1	42.8	7 31.3	57.1	4 36.1	57.9	15 07.4	45.3	22 04.9	20.3
3	22 54.6	13.8	16 45.8	43.6	7 08.4	57.3	4 59.2	57.6	15 25.4	44.7	22 12.8	19.3
4	22 48.8	14.9	16 28.3	44.3	6 45.5	57.6	5 22.1	57.4	15 43.2	44.0	22 20.3	18.3
5	22 42.7	16.0	16 10.4	45.0	6 22.4	57.8	5 45.2	57.2	16 00.7	43.4	22 27.4	17.4
6	22 36.0	+17.1	15 52.2	+45.7	5 59.2	+58.0	6 08.0	+56.9	16 17.9	+42.7	22 34.2	+16.4
7	22 29.0	18.3	15 33.8	46.4	5 36.0	58.2	6 30.7	56.6	16 34.8	42.0	22 40.5	15.4
8	22 21.4	19.4	15 15.2	47.0	5 12.7	58.4	6 53.3	56.3	16 51.5	41.3	22 46.5	14.4
9	22 13.5	20.5	14 56.2	47.6	4 49.3	58.5	7 15.7	56.0	17 07.9	40.6	22 52.0	13.4
10	22 05.1	21.5	14 37.1	48.3	4 25.9	58.7	7 38.1	55.7	17 23.9	39.9	22 57.2	12.4
11	21 56.2	+22.6	14 17.6	+48.9	4 02.4	+58.8	8 00.3	+55.4	17 39.7	+39.1	23 01.9	+11.4
12	21 47.0	23.7	13 58.0	49.4	3 38.9	58.9	8 22.4	55.0	17 55.3	38.4	23 06.3	10.4
13	21 37.3	24.7	13 38.1	50.0	3 15.3	59.0	8 44.3	54.7	18 10.5	37.6	23 10.2	9.3
14	21 27.2	25.6	13 18.0	50.6	2 51.7	59.1	9 06.1	54.3	18 25.4	36.9	23 13.7	8.3
15	21 16.7	26.8	12 57.6	51.1	2 28.0	59.1	9 27.8	53.9	18 40.0	36.1	23 16.8	7.3
16	21 05.8	+27.0	12 37.1	+51.6	2 04.4	+59.2	9 49.3	+53.2	18 54.2	+35.3	23 19.5	+6.3
17	20 54.5	28.3	12 16.4	52.1	1 40.7	59.2	10 10.6	53.1	19 08.2	34.5	23 21.8	5.2
18	20 42.8	29.0	11 55.4	52.6	1 17.0	59.3	10 31.7	52.7	19 21.8	33.7	23 23.7	4.2
19	20 30.7	30.8	11 34.3	53.0	0 53.3	59.3	10 52.7	52.3	19 35.2	32.9	23 25.2	3.2
20	20 18.2	31.7	11 13.0	53.5	0 29.5	59.3	11 13.5	51.8	19 48.1	32.0	23 26.3	2.1
21	20 05.0	+32.7	10 51.5	+53.9	S 0 05.8	+59.3	11 34.6	+51.3	20 00.8	+31.2	23 26.9	+1.1
22	19 52.0	33.6	10 29.9	54.4	N 0 17.9	59.2	11 54.6	50.9	20 13.1	30.3	23 27.2	+0.1
23	19 38.4	34.5	10 08.0	54.8	0 41.5	59.1	12 14.9	50.4	20 25.1	29.5	23 27.0	− 1.0
24	19 24.4	35.4	9 46.1	55.1	1 05.2	59.1	12 34.9	49.9	20 36.7	28.6	23 26.4	2.0
25	19 10.1	36.3	9 23.0	55.5	1 28.8	59.0	12 54.8	49.4	20 47.9	27.7	23 25.4	3.0
26	18 55.4	+37.2	9 01.7	+55.8	1 52.4	+58.9	13 14.4	+48.8	20 58.8	+26.8	23 24.0	− 4.1
27	18 40.3	38.1	8 39.3	56.2	2 16.0	58.8	13 33.8	48.3	21 09.4	25.9	23 22.1	5.1
28	18 24.9	38.9	S 8 16.7	+56.5	2 39.5	58.7	13 53.0	47.7	21 19.6	25.0	23 19.9	6.1
29	18 09.2	39.7	...	...	3 02.9	58.6	14 12.0	47.1	21 29.4	24.1	23 17.3	7.1
30	17 53.2	40.5	...	...	3 26.3	58.4	N 14 30.7	+46.5	21 38.8	23.1	N 23 14.2	− 8.2
31	S 17 36.8	+41.3	...	...	N 3 49.7	+58.3	...	...	N 21 47.9	+22.2	...	...

TABLE XV. — (1911 Continued.)

Day of month.	July. Declination.	Diff. for 1 hour.	August. Declination.	Diff. for 1 hour.	September. Declination.	Diff. for 1 hour.	October. Declination.	Diff. for 1 hour.	November. Declination.	Diff. for 1 hour.	December. Declination.	Diff. for 1 hour.
1	N 23° 10'.7	−9'.2	N 18° 15'.0	−37'.1	N 8° 36'.2	−54'.2	S 2° 51'.3	−58'.3	S 14° 10'.1	−48'.6	S 21° 41'.1	−24'.2
2	23 06.8	10.2	18 00.0	37.9	8 14.5	54.5	3 14.6	58.0	14 29.4	48.0	21 50.6	23.2
3	23 02.6	11.2	17 44.7	38.6	7 52.6	54.8	3 37.8	58.1	14 48.5	47.4	21 59.6	22.1
4	22 57.9	12.2	17 29.1	39.3	7 30.6	55.1	4 01.1	58.0	15 07.3	46.8	22 08.2	21.0
5	22 52.8	13.2	17 13.3	40.0	7 08.5	55.4	4 24.3	57.9	15 25.9	46.2	22 16.4	20.0
6	22 47.3	−14.2	16 57.1	−40.7	6 46.3	−55.7	5 47.4	−57.8	15 44.2	−45.5	22 24.2	−18.9
7	22 41.5	15.2	16 40.7	41.4	6 24.0	56.0	5 10.5	57.6	16 02.3	44.9	22 31.5	17.8
8	22 35.2	16.2	16 24.0	42.0	6 01.5	56.2	5 33.5	57.4	16 20.1	44.2	22 38.4	16.7
9	22 28.5	17.1	16 07.1	42.7	5 39.0	56.5	5 56.4	57.2	16 37.7	43.5	22 44.9	15.6
10	22 21.5	18.1	15 49.9	43.3	5 16.4	56.7	6 19.3	57.0	16 54.9	42.8	22 50.9	14.5
11	22 14.1	−19.1	15 33.4	−44.0	4 53.6	−56.9	6 42.0	−56.8	17 11.9	−42.1	22 56.4	−13.3
12	21 06.2	20.0	15 14.7	44.6	4 30.8	57.1	7 04.7	56.6	17 28.6	41.3	23 01.5	12.2
13	21 58.1	21.0	14 56.8	45.2	4 07.9	57.3	7 27.3	56.4	17 45.0	40.6	23 06.2	11.1
14	21 49.5	21.9	14 38.6	45.8	3 45.0	57.5	7 49.8	56.1	18 01.0	39.8	23 10.4	9.9
15	21 40.5	22.8	14 20.2	46.3	3 22.0	57.7	8 12.2	55.8	18 16.8	39.0	23 14.1	8.7
16	21 31.2	−23.7	14 01.5	−46.9	2 58.9	−57.8	8 34.5	−55.5	18 32.2	−38.2	23 17.4	−7.6
17	21 21.6	24.7	13 42.7	47.5	2 35.7	58.0	8 56.6	55.5	18 47.4	37.4	23 20.2	6.4
18	21 11.5	25.6	13 23.6	48.0	2 12.5	58.1	9 18.7	54.9	19 02.1	36.5	23 22.5	5.2
19	21 01.1	26.5	13 04.2	48.5	1 49.3	58.2	9 40.6	54.6	19 16.6	35.7	23 24.4	4.1
20	20 50.4	27.3	12 44.7	49.1	1 26.0	58.3	10 02.3	54.2	19 30.7	34.8	23 25.8	2.9
21	20 39.3	−28.2	12 25.0	−49.6	1 02.6	−58.4	10 24.0	−53.9	19 44.4	−33.9	23 26.7	−1.7
22	20 27.8	29.1	12 05.1	50.1	0 39.3	58.4	10 45.4	53.5	19 57.8	33.0	23 27.1	0.5
23	20 16.0	29.8	11 45.0	50.5	N 0 15.9	58.5	11 06.7	53.0	20 10.8	32.1	23 27.1	+ 0.6
24	20 03.8	30.8	11 24.7	51.0	S 0 07.5	58.5	11 27.8	52.6	20 23.5	31.1	23 26.6	1.8
25	19 51.4	31.6	11 04.2	51.4	0 30.9	58.5	11 48.8	52.2	20 35.7	30.2	23 25.6	3.0
26	19 38.6	−32.4	10 43.5	−51.9	0 54.3	−58.5	12 09.6	−51.7	20 47.6	−29.2	23 24.2	+ 4.2
27	19 25.4	33.3	10 22.7	52.3	1 17.7	58.5	12 30.2	51.2	20 59.1	28.2	23 22.3	5.4
28	19 12.0	34.1	10 01.7	52.7	1 41.1	58.5	12 50.6	50.7	21 10.2	27.3	23 19.9	6.5
29	18 58.2	34.8	09 40.5	53.1	S 2 04.5	58.5	13 10.8	50.2	21 20.9	26.2	23 17.1	7.7
30	18 44.1	35.6	09 19.2	53.5	S 2 27.9	58.4	S 13 30.7	49.7	S 21 31.2	−25.2	S 23 13.8	8.9
31	N 18 29.7	−36.4	N 8 57.8	−53.8	……	……	S 13 50.5	−49.2	……	……	S 23 10.0	+10.0

TABLE XV.—SUN'S DECLINATION AT GREENWICH MEAN NOON.
—1912—

Day of month	January Declination	January Diff. for 1 hour	February Declination	February Diff. for 1 hour	March Declination	March Diff. for 1 hour	April Declination	April Diff. for 1 hour	May Declination	May Diff. for 1 hour	June Declination	June Diff. for 1 hour
1	S 23° 05.7	+11".2	S 17° 24.1	+41".9	S 7° 36.7	+57".0	N 4° 30.7	+57".9	N 15° 03.1	+45".4	N 22° 02.9	+20'.5
2	23 01.0	12.3	17 07.2	42.6	7 13.8	57.2	4 53.8	57.6	15 21.2	44.8	22 10.9	19.5
3	22 55.9	13.5	16 50.0	43.4	6 50.9	57.5	5 16.8	57.4	15 38.9	44.2	22 18.5	18.6
4	22 50.3	14.6	16 32.5	44.1	6 27.9	57.7	5 39.7	57.2	15 56.5	43.5	22 25.7	17.6
5	22 44.2	15.7	16 14.7	44.8	6 04.7	57.9	6 02.5	56.9	16 13.7	42.8	22 32.6	16.6
6	22 37.7	+16.9	15 56.6	+45.5	5 41.5	+58.1	6 25.3	+56.7	16 30.8	+42.2	22 39.0	+15.6
7	22 30.7	18.0	15 38.3	46.2	5 18.3	58.3	6 47.9	56.4	16 47.5	41.5	22 45.1	14.6
8	22 23.3	19.1	15 19.7	46.9	4 54.9	58.5	7 10.4	56.1	17 03.9	40.8	22 50.7	13.6
9	22 15.4	20.2	15 00.8	47.5	4 31.5	58.6	7 32.7	55.8	17 20.1	40.1	22 56.0	12.6
10	22 07.1	21.3	14 41.7	48.1	4 08.0	58.8	7 55.0	55.5	17 36.0	39.3	23 00.8	11.6
11	21 58.4	+22.3	14 22.3	+48.7	3 44.5	+58.9	8 17.1	+55.1	17 51.6	+38.6	23 05.3	+10.6
12	21 49.3	23.4	14 02.7	49.3	3 20.9	59.0	8 39.1	54.8	18 06.9	37.9	23 09.3	9.6
13	21 39.7	24.5	13 42.8	49.9	2 57.3	59.1	9 01.0	54.4	18 21.8	37.1	23 12.9	8.6
14	21 29.7	25.5	13 22.8	50.6	2 33.6	59.2	9 22.7	54.1	18 36.5	36.3	23 16.2	7.5
15	21 19.3	26.6	13 02.5	51.0	2 10.0	59.2	9 44.2	53.7	18 50.9	35.5	23 19.0	6.5
16	21 08.5	+27.6	12 42.0	+51.5	1 46.2	+59.3	10 05.6	+53.3	19 04.9	+34.7	23 21.4	+ 5.5
17	20 57.2	28.6	12 21.3	52.0	1 22.5	59.3	10 26.8	52.8	19 18.7	33.9	23 23.4	4.5
18	20 45.6	29.6	12 00.4	52.5	0 58.8	59.3	10 47.8	52.4	19 32.1	33.1	23 24.9	3.4
19	20 33.6	30.6	11 39.3	53.0	S 0 35.1	59.3	11 08.7	51.9	19 45.1	32.2	23 26.4	2.4
20	20 21.2	31.5	11 18.0	53.4	S 0 11.4	59.3	11 29.4	51.5	19 57.9	31.4	23 26.8	1.3
21	20 08.4	+32.5	10 56.5	+53.9	N 0 12.4	+59.3	11 49.9	+51.0	20 10.3	+30.5	23 27.2	+ 0.3
22	19 55.2	33.4	10 34.9	54.3	0 36.1	59.2	12 10.2	50.5	20 22.3	29.7	23 27.1	0 —
23	19 41.6	34.3	10 13.1	54.7	0 59.7	59.1	12 30.3	50.0	20 34.0	28.8	23 26.6	1.8
24	19 27.7	35.2	9 51.2	55.0	1 23.4	59.1	12 50.2	49.5	20 45.3	27.9	23 25.7	2.8
25	19 13.5	36.1	9 29.1	55.4	1 47.0	59.0	13 09.8	48.9	20 56.3	27.0	23 24.3	3.8
26	18 58.8	+37.0	9 06.9	+55.7	2 10.5	+58.8	13 29.3	+48.4	21 06.9	+26.1	23 22.6	— 4.9
27	18 43.9	37.9	8 44.5	56.1	2 34.0	58.7	13 48.5	47.8	21 17.2	25.2	23 20.5	5.9
28	18 28.6	38.7	8 22.0	56.4	2 57.5	58.6	14 07.5	47.1	21 27.1	24.3	23 17.9	6.9
29	18 12.9	39.5	S 7 59.4	+56.7	3 20.9	58.4	14 26.3	46.6	21 36.6	23.3	23 15.0	7.9
30	17 57.0	40.3	……	……	3 44.2	58.2	N 14 44.8	+46.0	21 45.7	22.4	N 23 11.6	— 8.9
31	S 17 40.7	+41.1	……	……	N 4 07.5	+58.1	……	……	N 21 54.5	+21.4	……	……

TABLE XV. — (1912 Continued.)

Day of month	July Declination	July Diff. for 1 hour	August Declination	August Diff. for 1 hour	September Declination	September Diff. for 1 hour	October Declination	October Diff. for 1 hour	November Declination	November Diff. for 1 hour	December Declination	December Diff. for 1 hour
1	N 23° 07.8	−9.9	N 18° 03.7	−37.7	N 8° 10.8	−54.4	S 3° 08.8	−58.3	S 14° 24.6	−48.2	S 21° 48.3	−23.4
2	23 03.6	10.0	17 48.5	38.4	7 58.0	54.7	3 32.1	58.2	14 43.8	47.6	21 57.4	22.4
3	22 59.1	12.0	17 33.0	39.1	7 36.1	55.0	3 55.3	58.1	15 02.7	47.0	22 06.2	21.3
4	22 54.1	13.0	17 17.2	39.8	7 14.0	55.4	4 18.5	58.0	15 21.3	46.4	22 14.5	20.3
5	22 48.7	13.9	17 01.1	40.5	6 51.8	55.6	4 41.7	57.8	15 39.8	45.7	22 22.4	19.2
6	22 42.9	−14.9	16 44.8	−41.2	6 29.5	−55.9	5 04.8	−57.7	15 57.9	−45.1	22 29.8	−18.1
7	22 36.8	15.9	16 28.2	41.9	6 07.0	56.2	5 27.8	57.5	16 15.8	44.4	22 36.8	17.0
8	22 30.2	16.9	16 11.3	42.5	5 44.5	56.4	5 50.8	57.3	16 33.4	43.7	22 43.4	15.9
9	22 23.3	17.9	15 54.1	43.2	5 21.0	56.7	6 13.7	57.2	16 50.8	43.0	22 49.5	14.7
10	22 15.9	18.8	15 36.7	43.8	4 59.2	56.9	6 36.5	56.9	17 07.8	42.3	22 55.2	13.6
11	22 08.2	−19.8	15 19.1	−44.5	4 36.4	−57.1	6 59.3	−56.7	17 24.6	−41.5	23 00.4	−12.5
12	22 00.1	20.7	15 01.2	45.1	4 13.5	57.3	7 21.9	56.5	17 41.1	40.8	23 05.1	11.3
13	21 51.6	21.7	14 43.0	45.7	3 50.5	57.5	7 44.4	56.2	17 57.2	40.0	23 09.4	10.2
14	21 42.8	22.6	14 24.6	46.2	3 27.5	57.6	8 06.9	55.9	18 13.1	39.2	23 13.3	9.0
15	21 33.5	23.5	14 06.0	46.8	3 04.4	57.8	8 29.2	55.6	18 28.6	38.4	23 16.6	7.8
16	21 23.9	−24.5	13 47.2	−47.4	2 41.3	−57.9	8 51.3	−55.3	18 43.8	−37.6	23 19.5	−6.7
17	21 14.0	25.4	13 28.1	47.9	2 18.1	58.0	9 13.4	55.0	18 58.6	36.7	23 22.0	5.5
18	21 03.6	26.3	13 08.9	48.4	1 54.8	58.1	9 35.3	54.6	19 13.1	35.9	23 24.0	4.3
19	20 53.0	27.1	12 49.4	48.9	1 31.6	58.2	9 57.1	54.3	19 27.3	35.0	23 25.5	3.2
20	20 41.9	28.0	12 29.7	49.4	1 08.3	58.3	10 18.8	53.9	19 41.1	34.1	23 26.5	2.0
21	20 30.6	−28.9	12 09.8	−49.9	N 0 44.0	−58.4	S 10 40.2	−53.5	S 19 54.6	−33.2	S 23 27.1	−0.8
22	20 18.8	29.7	11 50.6	50.4	N 0 21.6	58.4	11 01.6	53.1	20 07.7	32.3	23 27.1	+0.4
23	20 06.8	30.6	11 30.6	50.9	S 0 01.8	58.5	11 22.7	52.7	S 20 20.4	31.4	23 26.8	1.5
24	19 54.4	31.4	11 09.1	51.3	S 0 25.2	58.5	11 43.7	52.2	20 32.8	30.4	23 25.9	2.7
25	19 41.7	32.2	10 48.5	51.7	0 48.6	58.5	12 04.5	51.8	20 44.7	29.4	23 24.6	3.9
26	19 28.6	−33.0	10 27.8	−52.1	S 1 12.0	−58.5	S 12 25.1	−51.2	20 56.3	−28.5	S 23 22.8	+5.1
27	19 15.2	33.8	10 06.8	52.6	1 35.4	58.5	12 45.6	50.8	S 21 07.5	27.5	23 20.6	6.2
28	19 01.6	34.6	9 45.7	52.9	1 58.8	58.4	13 05.8	50.3	21 18.3	26.5	23 17.8	7.4
29	18 47.6	35.4	9 24.5	53.3	S 2 22.1	58.4	13 25.8	49.8	21 28.7	25.5	23 14.6	8.6
30	18 33.2	36.2	9 03.1	53.7	2 45.5	58.3	13 45.6	49.3	S 21 38.7	24.5	23 11.0	9.7
31	N 18 18.6	−36.9	N 8 41.5	−54.0	.....	.....	S 14 05.2	−48.7	.....	.....	S 23 06.8	+10.9

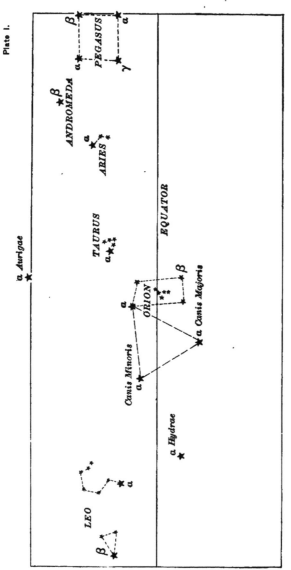

Plate I.

PRINCIPAL FIXED STARS BETWEEN 45° NORTH AND 45° SOUTH OF EQUATOR.

Plate II.

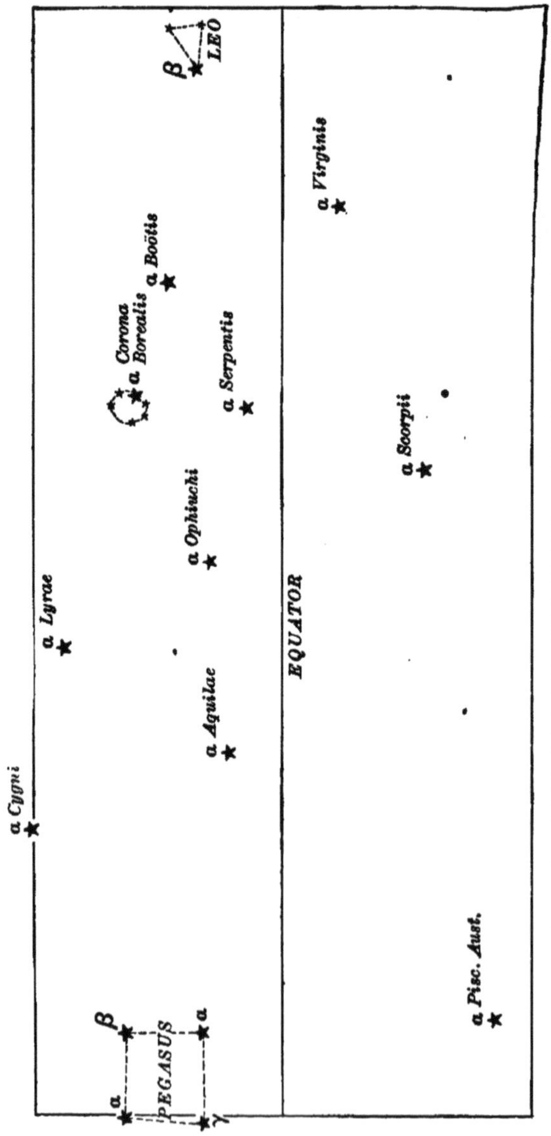

α Cygni

β α PEGASUS α γ

α Lyrae

α Aquilae

α Ophiuchi

Corona Borealis α Boötis

α Serpentis

α Virginis

LEO β

α Scorpii

EQUATOR

α Pisc. Aust.

PRINCIPAL FIXED STARS BETWEEN 45° NORTH AND 45° SOUTH OF EQUATOR.

MISCELLANEOUS RULES
AND TABLES

CORRECTING THE WATCH.

To find the approximate time, for correcting the sun's declination:

1. Add the log sin azimuth (Table XIII B) to the log cos altitude and from the sum subtract the log cos declination. The result is the log sin hour angle. Convert this angle into hours, minutes, and seconds. If it is forenoon, subtract this hour angle from 12^h to obtain the solar time.

2. Convert the time thus found into Mean Time by adding or subtracting the Equation of Time (Table XIII A).

3. Convert this Mean Time into Standard Time by taking the difference in longitude between the place and the standard meridian, expressing it as hours, minutes, and seconds, and adding it to the mean time if the place is west of the Standard Meridian, subtracting if it is east. The difference between this and the average watch reading is the error of the watch.*

This problem might be applied as follows. If the surveyor is far from a place where the time can be obtained he may work up his azimuth observation as previously explained and then compute the time by this method. If it is discovered that the watch is largely in error the azimuth should be recomputed, using the corrected declination of the sun. If the time is thus computed with each azimuth observation the error of the watch may be known approximately at all times.

The only doubtful case in the above solution is when the sun is about 6^h east or west of the meridian. The sine will be the same for the hour angle or its supplement. To remove this ambiguity compute the altitude of the sun when the hour angle is 6^h, which is done by adding the log sin latitude to the log sin declination. The result is the log sin altitude at the instant when the sun is 6^h from the meridian. If the observed altitude is less than this computed altitude the hour angle is greater than $90°$ and vice versa.

---

* If it is desired to compute the sun's hour angle without first computing the azimuth this may be done by means of equation [6].

EXAMPLE. From example 2, p. 12, we have Declination = + 21° 40', Altitude = 57° 20', Azimuth = S. 59° 55' E., Eastern Time 9ʰ 49ᵐ, A.M. The longitude is approximately 71° 01' W. Date, July 15.

log sin azimuth =	.938	Standard Meridian	75°
log cos altitude =	.732	Local Meridian	71° 01'
	1.670	Difference =	3° 59'
log cos declination =	.969	=	15ᵐ 56ˢ,
log sin hour angle =	.701		
hour angle =	30°.2		
=	2ʰ 01ᵐ		
Time =	9ʰ 59ᵐ		
Equation =	+6ᵐ		
Mean Time =	10ʰ 05ᵐ		
Longitude correction	16ᵐ		
Eastern Time	9ʰ 49ᵐ	showing that the watch was correct.	

In example 3, p. 12, the declination is + 16° 50'.7 and the latitude is 42° 29'.2.

log sin declination =	.462
log sin latitude =	.830
log sin altitude =	.292
Altitude at 6ʰ =	11°.3.

Hence the hour angle of the sun at the time of the observation was less than 6ʰ, since the observed altitude is greater than 11°.3.

TABLE XVI.—STADIA MEASUREMENTS.

Vert. Angle.	Difference in elevation for distance 100.						Proportional parts.								
	00'	10'	20'	30'	40'	50'	1'	2'	3'	4'	5'	6'	7'	8'	9'
0°	0.00	0.29	0.58	0.87	1.16	1.45	.03	.06	.09	.12	.15	.17	.20	.23	.26
1	1.74	2.04	2.33	2.62	2.91	3.20	.03	.06	.09	.12	.15	.17	.20	.23	.26
2	3.49	3.78	4.07	4.36	4.65	4.94	.03	.06	.09	.12	.15	.17	.20	.23	.26
3	5.23	5.52	5.80	6.09	6.38	6.67	.03	.06	.09	.12	.14	.17	.20	.23	.26
4	6.96	7.25	7.53	7.82	8.11	8.40	.03	.06	.09	.12	.14	.17	.20	.23	.26
5	8.68	8.97	9.25	9.54	9.83	10.11	.03	.06	.09	.11	.14	.17	.20	.23	.26
6	10.40	10.68	10.96	11.25	11.53	11.81	.03	.06	.08	.11	.14	.17	.20	.23	.25
7	12.10	12.38	12.66	12.94	13.22	13.50	.03	.06	.08	.11	.14	.17	.19	.22	.25
8	13.78	14.06	14.34	14.62	14.90	15.17	.03	.06	.08	.11	.14	.17	.19	.22	.25
9	15.45	15.73	16.00	16.28	16.55	16.83	.03	.06	.08	.11	.14	.17	.19	.22	.25
10	17.10	17.37	17.65	17.92	18.19	18.46	.03	.05	.08	.11	.14	.16	.19	.22	.24
11	18.73	19.00	19.27	19.54	19.80	20.07	.03	.05	.08	.11	.13	.16	.18	.21	.24
12	20.34	20.60	20.87	21.13	21.39	21.66	.03	.05	.08	.10	.13	.16	.18	.21	.24
13	21.92	22.18	22.44	22.70	22.96	23.22	.03	.05	.08	.10	.13	.16	.18	.21	.23
14	23.47	23.73	23.99	24.24	24.49	24.75	.03	.05	.08	.10	.13	.15	.18	.20	.23
15	25.00	25.25	25.50	25.75	26.00	26.25	.03	.05	.08	.10	.13	.15	.18	.20	.23
16	26.50	26.74	26.99	27.23	27.48	27.72	.02	.05	.07	.10	.12	.15	.17	.20	.22
17	27.96	28.20	28.44	28.68	28.92	29.15	.02	.05	.07	.10	.12	.14	.17	.19	.22
18	29.39	29.62	29.86	30.09	30.32	30.55	.02	.05	.07	.09	.12	.14	.16	.19	.21
19	30.78	31.01	31.24	31.47	31.69	31.92	.02	.05	.07	.09	.11	.14	.16	.18	.21
20	32.14	32.36	32.58	32.80	33.02	33.24	.02	.04	.07	.09	.11	.13	.16	.18	.20
21	33.46	33.67	33.89	34.10	34.31	34.52	.02	.04	.06	.08	.11	.13	.15	.17	.19
22	34.73	34.94	35.15	35.36	35.56	35.76	.02	.04	.06	.08	.10	.12	.14	.16	.18
23	35.97	36.17	36.37	36.57	36.77	36.96	.02	.04	.06	.08	.10	.12	.14	.16	.18
24	37.16	37.35	37.54	37.74	37.93	38.11	.02	.04	.06	.08	.10	.11	.13	.15	.17
25	38.30	38.49	38.67	38.86	39.04	39.22	.02	.04	.06	.07	.09	.11	.13	.15	.17
26	39.40	39.58	39.76	39.93	40.11	40.28	.02	.03	.05	.07	.09	.11	.12	.14	.16
27	40.45	40.62	40.79	40.96	41.12	41.29	.02	.03	.05	.07	.08	.10	.12	.13	.15
28	41.45	41.61	41.77	41.93	42.09	42.25	.02	.03	.05	.06	.08	.10	.11	.13	.14
29	42.40	42.56	42.71	42.86	43.01	43.16	.02	.03	.04	.06	.08	.09	.11	.12	.14

TABLE XVI. — (Continued).

Vert. Angle.	Horizontal corrections for distance 100.						Proportional parts.								
	00′	10′	20′	30′	40′	50′	1′	2′	3′	4′	5′	6′	7′	8′	9′
0°	0.00	0.00	0.00	0.01	0.01	0.02	…	…	…	…	…	…	…	…	…
1	0.03	0.04	0.05	0.07	0.08	0.10	.00	.00	.00	.00	.01	.01	.01	.01	.01
2	0.12	0.14	0.17	0.19	0.22	0.24	.00	.00	.01	.01	.01	.01	.01	.02	.02
3	0.27	0.31	0.34	0.37	0.41	0.44	.00	.01	.01	.01	.02	.02	.02	.02	.03
4	0.49	0.53	0.57	0.62	0.66	0.71	.00	.01	.01	.02	.02	.02	.03	.03	.04
5	0.76	0.81	0.86	0.92	0.97	1.03	.01	.01	.02	.02	.03	.03	.04	.04	.05
6	1.09	1.15	1.22	1.28	1.35	1.42	.01	.01	.02	.02	.03	.04	.04	.05	.05
7	1.49	1.56	1.63	1.71	1.78	1.86	.01	.01	.02	.03	.04	.04	.05	.06	.06
8	1.94	2.02	2.10	2.18	2.27	2.36	.01	.02	.02	.03	.04	.05	.06	.06	.07
9	2.45	2.54	2.63	2.72	2.82	2.92	.01	.02	.03	.04	.04	.05	.06	.07	.08
10	3.02	3.12	3.22	3.32	3.43	3.53	.01	.02	.03	.04	.05	.06	.07	.08	.09
11	3.64	3.75	3.86	3.97	4.09	4.21	.01	.02	.03	.04	.05	.07	.08	.09	.10
12	4.32	4.44	4.56	4.68	4.81	4.93	.01	.02	.04	.05	.06	.07	.08	.09	.11
13	5.06	5.19	5.33	5.45	5.58	5.72	.01	.03	.04	.05	.06	.08	.09	.10	.11
14	5.85	5.99	6.13	6.27	6.41	6.55	.01	.03	.04	.05	.07	.08	.10	.11	.12
15	6.70	6.84	6.99	7.14	7.29	7.44	.01	.03	.04	.06	.07	.09	.10	.12	.13
16	7.60	7.75	7.91	8.07	8.23	8.39	.02	.03	.05	.06	.08	.09	.11	.12	.14
17	8.55	8.71	8.88	9.04	9.21	9.38	.02	.03	.05	.07	.08	.10	.11	.13	.15
18	9.55	9.72	9.89	10.07	10.24	10.42	.02	.03	.05	.07	.09	.10	.12	.14	.15
19	10.60	10.78	10.96	11.14	11.33	11.51	.02	.04	.05	.07	.09	.11	.13	.14	.16
20	11.70	11.89	12.07	12.26	12.46	12.65	.02	.04	.06	.08	.10	.11	.13	.15	.17
21	12.84	13.04	13.23	13.43	13.63	13.83	.02	.04	.06	.08	.10	.12	.14	.16	.18
22	14.03	14.24	14.44	14.64	14.85	15.06	.02	.04	.06	.08	.10	.12	.14	.16	.18
23	15.27	15.48	15.69	15.90	16.11	16.33	.02	.04	.06	.08	.10	.13	.15	.17	.19
24	16.54	16.76	16.98	17.20	17.42	17.64	.02	.04	.06	.09	.11	.13	.15	.17	.19
25	17.86	18.08	18.31	18.53	18.76	18.99	.02	.04	.07	.09	.11	.14	.16	.18	.20
26	19.22	19.45	19.68	19.91	20.14	20.38	.02	.05	.07	.09	.12	.14	.16	.19	.21
27	20.61	20.85	21.08	21.32	21.56	21.80	.02	.05	.07	.10	.12	.14	.17	.19	.22
28	22.04	22.28	22.52	22.77	23.01	23.26	.03	.05	.08	.10	.12	.14	.17	.19	.22
29	23.50	23.75	24.00	24.25	24.50	24.75	.03	.05	.08	.10	.12	.15	.17	.20	.23

TABLE XVII. — CORRECTIONS FOR REDUCING SLOPE MEASUREMENTS TO HORIZONTAL.

Vert. Angle	ft. 100	ft. 200	ft. 300	Feet. 10	20	30	40	50	60	70	80	90
0° 30′	.00	.01	.01	......	......	......	......	......	......	....	....	....
40′	.01	.01	.02	......	......	......	......	......	......	.01	.01	.01
50	.01	.02	.03	......	......	......	......	.01	.01	.01	.01	.01
1° 00′	.02	.03	.05	......	......	......	.01	.01	.01	.01	.01	.01
10′	.02	.04	.06	......	......	.01	.01	.01	.01	.01	.02	.02
20′	.03	.05	.08	......	.01	.01	.01	.01	.02	.02	.02	.02
30′	.03	.07	.10	......	.01	.01	.01	.02	.02	.02	.03	.03
40′	.04	.08	.13	......	.01	.01	.02	.02	.03	.03	.03	.04
50′	.05	.10	.15	.01	.01	.02	.02	.03	.03	.04	.04	.05
2° 00′	.06	.12	.18	.01	.01	.02	.02	.03	.04	.04	.05	.05
10′	.07	.14	.21	.01	.01	.02	.03	.04	.04	.05	.06	.06
20′	.08	.17	.25	.01	.02	.02	.03	.04	.05	.06	.07	.07
30′	.10	.19	.29	.01	.02	.03	.04	.05	.06	.07	.08	.09
40′	.11	.22	.32	.01	.02	.03	.04	.05	.06	.08	.09	.10
50′	.12	.24	.37	.01	.02	.04	.05	.06	.07	.09	.10	.11
3° 00′	.14	.27	.41	.01	.03	.04	.05	.07	.08	.10	.11	.12
10′	.15	.31	.46	.02	.03	.05	.06	.08	.09	.11	.12	.14
20′	.17	.34	.51	.02	.03	.05	.07	.08	.10	.12	.14	.15
30′	.19	.37	.56	.02	.04	.06	.07	.09	.11	.13	.15	.17
40′	.20	.41	.61	.02	.04	.06	.08	.10	.12	.14	.16	.18
50′	.22	.45	.67	.02	.04	.07	.09	.11	.13	.16	.18	.20
4° 00′	.24	.49	.73	.02	.05	.07	.10	.12	.15	.17	.19	.22
10′	.26	.53	.79	.03	.05	.08	.11	.13	.16	.19	.21	.24
20′	.29	.57	.86	.03	.06	.09	.11	.14	.17	.20	.23	.26
30′	.31	.62	.92	.03	.06	.09	.12	.15	.18	.22	.25	.28
40′	.33	.66	.99	.03	.07	.10	.13	.17	.20	.23	.27	.30
50′	.36	.71	1.07	.04	.07	.11	.14	.18	.21	.25	.28	.32
5° 00′	.38	.76	1.14	.04	.08	.11	.15	.19	.23	.27	.30	.34
10′	.41	.81	1.22	.04	.08	.12	.16	.20	.24	.28	.33	.37
20′	.43	.87	1.30	.04	.09	.13	.17	.22	.26	.30	.35	.39
30′	.46	.92	1.38	.05	.09	.14	.18	.23	.28	.32	.37	.41
40′	.49	.98	1.47	.05	.10	.15	.20	.24	.29	.34	.39	.44
50′	.52	1.04	1.55	.05	.10	.16	.21	.26	.31	.36	.41	.47
6° 00′	.55	1.10	1.64	.05	.11	.16	.22	.27	.33	.38	.44	.49
10′	.58	1.16	1.74	.05	.12	.17	.23	.29	.35	.41	.46	.52
20′	.61	1.22	1.83	.06	.12	.18	.24	.31	.37	.43	.49	.55
30′	.64	1.29	1.93	.06	.13	.19	.26	.32	.39	.45	.51	.58
40′	.68	1.35	2.03	.07	.14	.20	.27	.34	.41	.47	.54	.61
50′	.71	1.42	2.13	.07	.14	.21	.28	.36	.43	.50	.57	.64

EXAMPLE. — Required the horizontal distance for a slope distance of 272.46 ft., vertical angle of 4° 16′. For 4° 10′ the correction is 0.53 + 0.19 = 0 72 ft. Interpolating between 4° 10′ and 4° 20′ for distance 300 we find 0.04 as the increase for 6′. Hence correction = 0.76, and horizontal distance = 271.70 ft.

TABLE XVIII. — INCHES IN DECIMALS OF A FOOT.

In.	0	1	2	3	4	5	6	7	8	9	10	11	In.
0	Feet	.0833	.1667	.2500	.3333	.4167	.5000	.5833	.6667	.7500	.8333	.9167	0
1/32	.0026	.0859	.1693	.2526	.3359	.4193	.5026	.5859	.6693	.7526	.8359	.9193	1/32
1/16	.0052	.0885	.1719	.2552	.3385	.4219	.5052	.5885	.6719	.7552	.8385	.9219	1/16
3/32	.0078	.0911	.1745	.2578	.3411	.4245	.5078	.5911	.6745	.7578	.8411	.9245	3/32
1/8	.0104	.0938	.1771	.2604	.3438	.4271	.5104	.5938	.6771	.7604	.8438	.9271	1/8
5/32	.0130	.0964	.1797	.2630	.3464	.4297	.5130	.5964	.6797	.7630	.8464	.9297	5/32
3/16	.0156	.0990	.1823	.2656	.3490	.4323	.5156	.5990	.6823	.7656	.8490	.9323	3/16
7/32	.0182	.1016	.1849	.2682	.3516	.4349	.5182	.6016	.6849	.7682	.8516	.9349	7/32
1/4	.0208	.1042	.1875	.2708	.3542	.4375	.5208	.6042	.6875	.7708	.8542	.9375	1/4
9/32	.0234	.1068	.1901	.2734	.3568	.4401	.5234	.6068	.6901	.7734	.8568	.9401	9/32
5/16	.0260	.1094	.1927	.2760	.3594	.4427	.5260	.6094	.6927	.7760	.8594	.9427	5/16
11/32	.0286	.1120	.1953	.2786	.3620	.4453	.5286	.6120	.6953	.7786	.8620	.9453	11/32
3/8	.0313	.1146	.1979	.2813	.3646	.4479	.5313	.6146	.6979	.7813	.8646	.9479	3/8
13/32	.0339	.1172	.2005	.2839	.3672	.4505	.5339	.6172	.7005	.7839	.8672	.9505	13/32
7/16	.0365	.1198	.2031	.2865	.3698	.4531	.5365	.6198	.7031	.7865	.8698	.9531	7/16
15/32	.0391	.1224	.2057	.2891	.3724	.4557	.5391	.6224	.7057	.7891	.8724	.9557	15/32
1/2	.0417	.1250	.2083	.2917	.3750	.4583	.5417	.6250	.7083	.7917	.8750	.9583	1/2
17/32	.0443	.1276	.2109	.2943	.3776	.4609	.5443	.6276	.7109	.7943	.8776	.9609	17/32
9/16	.0469	.1302	.2135	.2969	.3802	.4635	.5469	.6302	.7135	.7969	.8802	.9635	9/16
19/32	.0495	.1328	.2161	.2995	.3828	.4661	.5495	.6328	.7161	.7995	.8828	.9661	19/32
5/8	.0521	.1354	.2188	.3021	.3854	.4688	.5521	.6354	.7188	.8021	.8854	.9688	5/8
21/32	.0547	.1380	.2214	.3047	.3880	.4714	.5547	.6380	.7214	.8047	.8880	.9714	21/32
11/16	.0573	.1406	.2240	.3073	.3906	.4740	.5573	.6406	.7240	.8073	.8906	.9740	11/16
23/32	.0599	.1432	.2266	.3099	.3932	.4766	.5599	.6432	.7266	.8099	.8932	.9766	23/32
3/4	.0625	.1458	.2292	.3125	.3958	.4792	.5625	.6458	.7292	.8125	.8958	.9792	3/4
25/32	.0651	.1484	.2318	.3151	.3984	.4818	.5651	.6484	.7318	.8151	.8984	.9818	25/32
13/16	.0677	.1510	.2344	.3177	.4010	.4844	.5677	.6510	.7344	.8177	.9010	.9844	13/16
27/32	.0703	.1536	.2370	.3203	.4036	.4870	.5703	.6536	.7370	.8203	.9036	.9870	27/32
7/8	.0729	.1563	.2396	.3229	.4063	.4896	.5729	.6563	.7396	.8229	.9063	.9896	7/8
29/32	.0755	.1589	.2422	.3255	.4089	.4922	.5755	.6589	.7422	.8255	.9089	.9922	29/32
15/16	.0781	.1615	.2448	.3281	.4115	.4948	.5781	.6615	.7448	.8281	.9115	.9948	15/16
31/32	.0807	.1641	.2474	.3307	.4141	.4974	.5807	.6641	.7474	.8307	.9141	.9974	31/32
In.	0	1	2	3	4	5	6	7	8	9	10	11	In.

Milton Keynes UK
Ingram Content Group UK Ltd.
UKHW020631171024
2225UKWH00053B/533